YOSSI CHALAMISH, M.D.

THE
BRAIN
CODE

Using Neuroscience to Improve
Learning, Memory and Emotional Intelligence

WATKINS
Sharing Wisdom

The Brain Code
Yossi Chalamish M.D.

First published in the UK and USA in 2024 by
Watkins, an imprint of Watkins Media Limited
Unit 11, Shepperton House, 89–93 Shepperton Road
London N1 3DF

enquiries@watkinspublishing.com

Publisher: Fiona Robertson
Commissioning Editor: Etan Ilfeld
Project Editor: Brittany Willis
Translator: Christopher Slaney
Copy Editor: Michelle Clark
Head of Design: Karen Smith
Production: Uzma Taj

A CIP record for this book is available from the British Library

ISBN: 978-1-78678-881-8 (Hardback)
ISBN: 978-1-78678-882-5 (eBook)

10 9 8 7 6 5 4 3 2 1

Typeset by Lapiz
Printed in the United Kingdom by TJ Books Ltd

Note: Many examples are given and stories told throughout, but the
people mentioned are either fictional or their names have been changed
to protect their identities, out of respect for their privacy.

www.watkinspublishing.com

MIX
Paper from
responsible sources
FSC
www.fsc.org   FSC® C013056

With love to my parents, Shimon
(of blessed memory) and Flora
Also to Maayan Ziv

CONTENTS

PREFACE

I am eager to share how you can help your brain function better by using its own algorithm – the brain code – to improve your memory, mental and emotional health, relationships, sleep and general wellbeing.

Let's start from the beginning of my journey. After completing my medical studies at one of Israel's premier universities devoted to scientific research, Technion, I chose to specialize in psychiatry. The human mind has always aroused my curiosity and I wanted to understand it better. Luck was on my side, and I was soon heading to the Department of Brain Research at the Weizmann Institute in Rehovot, Israel, to commence my postdoctoral research.

I didn't know it then, but this was a great time to be doing such work as it was the beginning of a golden age in brain research. Groundbreaking studies had discovered that the human brain can change continually, adapting to its environment, improving and even repairing itself. The term "the flexible brain" was born, and in the corridors of the Weizmann Institute (and in the cafeteria) there was huge excitement about the wondrous flexible brain, with people asking, "What does this mean?"

Over two decades have passed since those heady early discoveries of the flexible brain. And many neuroscientists today are grappling with the question of how to harness that understanding to improve our physical and mental health. I think that there are two main reasons why the theory of the flexible brain has not yet reinvigorated and changed how we practise medicine.

The first is that relatively few doctors are involved in brain research. This field has instead attracted biologists, programmers, engineers, physicists and mathematicians. As a result, more resources have been allocated to the practical and intriguing field of "brain–machine interfaces" and left behind the more intriguing study of what I shall term "flexible neurology".

Secondly, there is a tendency in Western society for scientists to specialize, as well as narrow, their areas of study over time. This tendency guarantees a laser-sharp focus on the chosen subject but prevents researchers from being able to zoom out and see the full picture. Naturally, this phenomenon extends to medicine. In the past, most doctors were generalists, their work varying greatly from treating diseases and wounds to obstetrics and ophthalmology. However, today, fields of specialism are becoming even narrower, such that, for example, in ophthalmology there are further subspecialties in the various parts of the eye. The situation is similar in other branches of medicine.

Brain research is no exception and, indeed, it is more evident in this field than in many others. The complexity of the human brain and all the knowledge we have about it mean that there are many areas to study: awareness, the brain–machine interface, the functions of emotion and creativity, and language and memory, to name but a few. As a result, a researcher focused on one of these areas will not be well acquainted with the others. Thus, a neuroscientist working on memory will only focus on the neural pathways in the brain and the unique structure of the brain in the areas associated with memory. They will also learn about the relay race-type mechanism that happens throughout the brain to enable the efficient retrieval of memories. Nevertheless, even while knowing all these things, they might be surprised to hear that walking improves memory, whereas excess sleep (not only a lack of it) damages our ability to remember things.

A possible solution to this situation was shown to me a few years ago when I taught a course in neuroanatomy at the Weizmann Institute. I was criticized by some of the students for paying too much attention to the big picture and not to the details. While initially these comments stung because I thought that I had created an exceptional course, it was a defining moment. In the years that followed, I took advantage of being both a doctor and a neuroscientist, combining my scientific knowledge with clinical experience to promote a new and fascinating scientific field: flexible brain therapy.

When my efforts in this field began to mature into this book, I decided to concentrate on topics that interest us all in our daily lives – from physical and mental health to happiness, memory and learning. I gathered knowledge and scientific data about them, processed it and distilled what I learned here so everyone can understand how the brain works. You will also find tools that I have derived from scientifically proven methods to apply to your daily life to improve how your brain functions.

The result? In each chapter of this book, you will meet a brain function that is intertwined with your everyday life. First, you can get to know it, the two of you becoming acquainted and even friends. Then, later, I describe how each function can be improved, so you can decide whether to settle for merely being acquainted or take a more active approach and journey toward self-improvement. However, there is no need to wait until the end of the book to start working as there are some practical exercises given at the end of each chapter to keep your brain actively engaged and continually improving.

Thanks to the understanding and knowledge you will gain from reading this book, you will be able to strengthen your immunity to diseases (such as flu and mild colds) and turn crises into opportunities. Included are physical and mental activities that you can integrate into your daily life and apply as needed to reduce mental stress. Studies prove that an active life that is as stress-free as possible promotes physical and mental health and longevity.

Wishing you good health,

Dr Yossi Chalamish

CHAPTER 1
A BRIEF HISTORY OF THE BRAIN

If I asked you to think about the human brain, you would probably conjure up an image of a rounded mass with a jelly-like texture and a surface like a walnut, with wiggly ridges and grooves. That's pretty accurate, but it's just the tip of the iceberg. Let's delve a little deeper.

The brain contains about 100 billion neurons – long, wiry structures adapted to transmit electrical signals to one another. Each neuron is made up of a cell body – the perikaryon – and connections that branch out of it – called axons – linking it to other neurons. The resulting structure is one huge neural network that has trillions of connections. This network allows us to move, feel, think, remember, be focused and stay motivated. It also supervises the activity of the body, enabling us to perform any one of thousands of actions at any given moment.

The activity of the brain takes place in four channels:

- **motor** – moving the skeleton
- **sensory** – processing sensory information
- **cognitive** – mental functions
- **supervisory** – overseeing the operation of all bodily systems, the scientific term for this being visceral homoeostasis

How the brain evolved

The first living creatures to have brains were fish. In the course of evolutionary development, creatures have only accumulated additional brain parts; they have not discarded areas once used by their ancestors. In other words, everything that existed, and still exists, in the fish brain also exists in our brain today. But, of course, the human brain contains many other large areas that the fish brain still does not.

The most ancient part of the human brain is the fish brain, so it is often referred to as the deep brain or subcortex. Some researchers see it as separate and claim that the human brain consists of two brains: the outer brain (the cerebral cortex or simple cortex) and this deep brain (the subcortex).

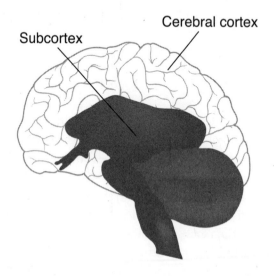

All brains were created to fulfil one purpose: to promote survival, of individuals and their descendants. This is what the fish brain – our deep brain – does, by trying to ensure that we stay alive in the here and now, so it does not need what we understand as "awareness". At every moment, our fish brain receives thousands of messages in the form of stimuli from the external environment (the world around us) and our internal environment (our body). These stimuli are received and processed into behavioural impulses that enable us to survive.

Later in the process of evolution, starting with turtles, a new brain area developed. This is the outermost layer that wraps around the deep brain – the cerebral cortex or cortex. The cortex expands the role of the brain beyond survival in the present to survival in the future, whether that be a day, a week, a year or more from now. It has one especially important ability: it picks up both the stimuli being received and the resulting behaviour.

During the evolutionary process, the human cortex grew until it became the broadest and most developed cortex in the animal kingdom. It grew so much that it could no longer fit into the human skull. That is why there are all those wiggly ridges and folds, as they greatly increase the surface area of the brain in the space available inside the skull. This enlarged cortex has allowed us to develop unique brain functions, such as language, a high level of awareness, the ability to act in ways that have deep meaning and, of course, the ability to learn new skills.

Thanks to rich and varied stimuli from the external world and our own bodies, processed by the massive neural network in the cortex, we have thoughts and impulses aimed at not only helping us to survive and thrive today but in the long term too.

The brain code

Brain researchers have discovered that disconnections and connections in our neural network continue to be made throughout our life.[1] The structure and functioning of the brain can change over time. Alterations occur when the brain has a basic understanding that some changes in the neural network may be useful for our immediate or long-term survival.

What is this basic understanding? It is helpful to think of it as an algorithm – a "brain code". Our brain code has been written and edited for millions of years and it enables the brain to interpret the myriad light and dark nuances of what's important and how we should behave in any given situation.

The brain code is a wonderful invention, but there's a problem. It was written millions of years ago and, since then, there have been major changes in the environment and how we live. Today, relatively, we live a life of abundance. Fierce animals are not trying to hunt us down, but we might feel threatened by an increase in our bills, a demanding boss or the possibility of being made redundant.

Does this mean that we are at the mercy of an algorithm that is fundamentally suited to our ancient ancestors? Definitely not. Although the brain code, which is the framework for brain activity, is not really up to date, we do enjoy the ability to change our brain patterns and adapt them to our needs. This is a natural and permanent ability, but for us to apply it, we must first become acquainted with the brain, its environment and its operating system.

CHAPTER 2
UNDERSTANDING MEMORY

At the end of a lecture, on the way to the car park, a silver-haired gentleman who had been sitting in the audience approached me. He was scared by what was happening to some of his friends as they aged, so he had been monitoring his memory and was worried about what he found. I asked him what kinds of things he had noticed, and he gave me this example:

> A few months ago, I saw a play at the theatre and really enjoyed it. For the past two days, I've been trying to remember the name of the play. I've racked my brain but am still drawing a blank. The name has simply been erased from my memory. Something must be out of kilter in my brain.

I was able to reassure him that he had nothing to worry about, his brain was not screwed up.

It has become clear to researchers that one of the good properties the brain has is forgetting details that do not help us to survive. Yadin Dudai, one of the leading scientists undertaking brain research, defines it like this: "The main goal of our memory is not the preservation of the past but the future."[1]

Therefore, I told the gentleman that the title of the play, no matter how much he had enjoyed it, would not contribute to his staying alive in the future, so his brain had simply deleted it. Our brain usually knows which memories to keep and which to forget. Holding on to excess memories would be an unnecessary burden that could disrupt our daily life. Controlled forgetting is essential if we are to function properly.

Seeing the sceptical look the gentleman gave me when I told him this, I looked for another way to convey the idea.

The example of date growers came to mind. The growers usually cut off palm fronds that seem unnecessary to them. That way, the fewer remaining fronds get a bigger share of the energy – water, nutrients and minerals – than they did before and so the quality of the dates produced improves.

The man still looked unconvinced, but it was late, so we both hurried on our way. As I saw him go, I could tell that I hadn't managed to get my idea across to him.

You, along with many others, may have found yourself in a similar situation, troubled about your inability to remember something that there was seemingly no reason to forget. To better understand and adopt tools to deal with forgetting, let's dive together into the depths of memory and, after deciphering its secrets, try to improve your memory. Before we do this, let's define what memory is.

What is memory?

Memory is the ability of living beings to receive messages from the environment, preserve them and use when needed. Every living being capable of survival and reproduction has an ability to remember even ones without a brain.

Researchers in Israel and Spain discovered that even amoebas (single-celled organisms) are endowed with the ability to remember.[2] Their findings reinforced the results of classic studies carried out in the 20th century.[3]

If amoebas can remember, creatures blessed with brains definitely can. Our sensory organs receive stimuli from the environment (from our senses of sight, hearing, taste, smell, touch) and transfer them as messages to the neural networks of memory found in the brain. In the case of fish, these networks are found exclusively in the subcortex, and their memories are not conscious. Whereas in humans the networks are found in both the subcortex and the cortex, and our memories are divided into unconscious and conscious ones.

Unconscious memory

To better understand the virtues of conscious memory and ways to improve it, we first need to deal with those messages that reach the neural systems in the subcortex without our being aware of them. They fly beneath our radar and go on to create unconscious memories. These memories are also retrieved when needed and can be seen in impulses that trigger our actions.

To illustrate how unconscious memory creates an impulse, I will share with you a visit I made to a relative who has an aquarium, full of magnificent fish. When I walked up to it, the fish continued to swim back and forth, completely ignoring my presence. However, when my relative – who feeds the fish every morning – came close, their reaction was different. The fish quickly swam toward him and opened their mouths in anticipation. His approach activated an unconscious memory that the fish shared. The stimulus of his image, received and processed by their eyes, passed a message to their memory networks in their subcortex, and because this message was familiar to their brain, the urge to come was triggered, as they knew from previous times that there could be food. All this happened unconsciously as fish do not have a cortex, which they would need to analyse this event, they simply follow their impulse.

Intuition or gut feeling?

Humans have the same network of unconscious memory that we recognize as intuition. Contrary to popular opinion, intuition is not simply a gut feeling that pushes us to act in pursuit of pleasure or to counter a threat. It is an urge that stems from a real memory where we learned new information that lodges in areas in the deep brain without us being aware that this process has happened. One of the leading scientists studying unconscious brain activity is David Eagleman, whose wonderful book *Incognito*[4] implies that, unlike a gut feeling, intuition is conditioned by rules and life experiences. The following example, taken from my own experience, will go some way toward explaining what intuition is.

For me, making the transition from writing with a pen to typing using a keyboard was not easy but it was necessary for my work. At first, I had to look at every letter on the keyboard; I couldn't type and think about the text I was writing at the same time. Some 15 years later, I am typing quickly without looking for the right keys, albeit still with only a couple of fingers. However, if you were to ask me what order the letters are in on the keyboard, I haven't a clue.

Why is this? Because the way I type relies on intuitive memory. Repeated use of the keyboard means that a certain area of my subcortex has come to know exactly where each letter is located. When I want to type a specific letter, my brain uses a link that has been made between that memory area and the areas that allow the movement of my fingers (the motor areas in the brain) and the relevant muscles move the correct finger to the key of the letter I need.

I claim that this is an intuitive action, but let's check if it is according to the two conditions test put forward by Eagleman. The two conditions are the existence of rules and life experience. In the typing example, rules exist: each letter has a fixed place on the keyboard. If the order of the letters for any one language was not the same on all keyboards used to type it, we would not be able to switch from one computer to another.

What of the life experience condition? Despite the letters being in fixed positions, it took me a long time to apply the right intuition to type the correct letters each time.

In contrast to intuition, which stems from unconscious memory, gut feelings are based only on guesswork. This is a kind of coin toss, a heads or tails guess, in which the chances of calling it correctly are even.

Now that we have the necessary tools to distinguish between intuition and gut feeling, let's consider the following scenario. Jayden , who is 35, visits his mother and, as soon as he opens the door, she takes one look at him and asks, "Did something happen?"

Jayden hurries to reassure her, replying, "No, Mother, everything is fine." But she insists: "I can tell when something's wrong. Tell me what happened."

If we were to ask Jayden's mother why she thinks something is wrong, she would probably answer, "I don't know, I have a gut feeling . . ."

Is she right? To answer that question, we need to check whether both Eagleman's conditions are met here. Jayden's statement, "No, Mother, everything is fine", might or might not be an honest indication of his mood. We all know that nuances of intonation, facial expressions and body language offer a clearer picture than words alone. These variables usually have their own rules and continuity from one situation to another. Each of us has our own personal body language with individual subtle tells that indicate whether we are lying or distressed about an emotional situation. Jayden's mother has known her son for 35 years. During his lifetime, she has learned to recognize such details of his body language and so she knows when he is not telling the truth he knits his eyebrows. This message was received multiple times over the years and lodged in the unconscious memory areas of her brain. Therefore, when her eyes catch sight of his eyebrows in such positions, an impulse based on her life experience tells her that something is not right, despite what he is saying. This is intuition, not a gut feeling.

Here is another example. The value of a company's stock suddenly drops by 8 per cent. Everyone knows this to be a stable and reputable company, so the drop seems illogical and perhaps the result of a mistake. There's an opportunity to buy the stock at a discounted price before it recovers. Would our decision to buy stock be based on intuition or gut feeling? From our experience of life, we know that stock prices go up and down, but we also know that there are no rules in this matter. The conclusion: it would be the result of a gut feeling.

Conscious memory

We will now climb from the ground floor of the brain– the subcortex – to the upstairs – the cortex – where we encounter conscious memory.

Declarative and procedural memory

Messages that reach the cortex become conscious memories, of which there are two types. The first is what we associate with names, places, the books we read, films and TV programmes we see, visits to the theatre and exhibitions, as well as where we left our keys, phone, glasses and other essentials. To express such a memory, we need language, so it is called declarative memory.

Alongside the neural network responsible for the declarative category of memories, there is another neural network of conscious memories that operates in the cortex related to things we see, hear and smell. As these are not accompanied by a verbal translation, this category is known as procedural memory.

Here is an example that will clarify the difference between declarative and procedural memory. If we see a certain house and say to ourselves, "Here is a house", this information will be recorded in the declarative network. But if we see the house and do not create a verbal translation of what we have seen, this event will find its place in the parallel neural network as a procedural memory. This also happens in cases of music, smells and movements that we do not define for ourselves. What these two categories have in common is that they both involve conscious exposure to stimuli from our environment.

Declarative memory is the younger sibling of procedural memory because it has only existed since we started to use language, about 70,000 years ago. When this happened, changes in the structure of the cortex and the anatomy of the tongue and throat further enabled speech. In the millions of years before this happened, humans did not know how to speak and the only type of conscious memory they had was procedural memory. When a tribe wandered from place to place in search of sustenance, its members needed to know where the nearest water was, the animals that could be hunted and where they could forage for ripe fruits and other foods to eat. Visual, auditory and olfactory memory were therefore sufficient for survival at that time. To be hunters, primitive people also needed a motor memory, which allowed them to throw a spear properly. All these attributes fall into the category of procedural memories.

As we have already learned, the mind does not sanctify the memory itself, it sanctifies survival and all that entails. Procedural memory proved its ability to promote everything related to survival long before the arrival of its younger sibling. Even today, the brain nurtures and favours the older, speechless sibling.

Here is a short story to help us understand how this works. Danny is six years old when his mother teaches him to ride his bike in the park. During the lesson, they meet Ruth, who is also learning to ride. The two children practise side by side and have fun. After they master the required skill and play together on the slides and swings, the two children go their separate ways.

The years pass and Danny abandons cycling. About 20 years later on a trip to Thailand, he wants to rent a bike. He is afraid that he won't be able to ride it, but decides to give it a go. To his great surprise, after a few minutes of wobbling, he gets the hang of it, as though it were only yesterday when he last went for a ride. In a similar way to our ancestors' spear-throwing skills, how to ride a bike is stored in the procedural network. As memories of this type are closely related to promoting survival, the memory of how to ride a bike is preserved well and, even 20 years later, can be retrieved without difficulty when needed. However, if you were to ask Danny for the name of the girl with whom he learned to ride, it is highly doubtful that he would be able to come up with it. According to the brain code, the preservation of the name, a declarative memory, is not necessary for Danny's survival, so the brain does not invest energy in it.

Failures in recalling declarative memories may simply constitute a natural phenomenon and not indicate any impairment of memory.

In this context, it is worth noting that, in the first stages of Alzheimer's disease, our procedural memory is not damaged because the brain attaches more importance to it than it does to declarative memories. That is why those with Alzheimer's disease will likely remember that they need to eat but may forget the name of the person who is serving them their meal.

Some tips for improving your memory

We will keep discussing declarative memory for the moment, to learn how it can be improved so that we can get better at remembering people's names, birthdays and other important details, such as where we put our keys.

The creation of a declarative memory and its use involves three stages:

1 **Encoding** – receiving a message via one of the senses that is also either spoken or we verbalize in some way and it goes into short-term memory
2 **Storage** – keeping the received message in a neural network in the cortex to become a long-term memory
3 **Retrieval** – using the stored message when it's needed

It often happens that the first two stages are performed correctly, but not the final and most important stage. Suppose I tell you that I recently visited a village in the north of the country. You show interest and ask me for the name of the village. I remember the sign at the entrance – "Welcome to . . ." – but the name of the village has faded from my memory. You try to help me and reel off names of villages in the north. I continue responding in the negative until you happen to mention the right name.

What happened here? The memory clearly exists. It successfully passed the storage phase and is now in the proper network, but I didn't have the ability to retrieve it. Nevertheless, when you suggested the right name, I immediately knew it was the one because the memory itself was properly stored and then I could retrieve it without difficulty. It was a one-time failure – the kind we all experience from time to time. If I had been presented with a list of three or four villages, including the right one, and I hadn't been able to spot it, *that* would be a cause for concern.

Improving declarative memory can be achieved on two levels: through the brain's hardware and software. The communication between the nerve cells is in the realm of the software (we will look at that later in this chapter). To start, let's look at the hardware and some simple steps we can take to improve it. As we have learned already, our brain is a large and complicated system, consisting of an unimaginable number of interconnected neurons called nerve cells.

Neurons are made in the body using a variety of chemical substances, such as proteins, sugars, fats, vitamins and minerals. Maintaining a supply of all these nutrients is key to ensuring the integrity of the brain's hardware and, therefore, of a good memory. It follows that if a person has an excellent memory, but suffers from a significant lack of omega 3, they would suffer an inability to remember basic information. This is similar to a high-end computer when one of its components has corroded. Its software is up to date, but the faulty hardware has rendered it unusable. Therefore, if a person (and it doesn't matter what age they are) came to me and complained about memory loss, the possibility of Alzheimer's disease is not the first thing I would consider. Instead, I would send them for a blood test.

The importance of a proper diet is clear. If possible, I would also advise everyone to schedule an appointment with a clinical dietician and tell them what you eat and drink each day. A simple computerized health screening blood test will indicate whether, depending on factors such as age and weight, you are giving your body and brain all the nutrition it needs. Studies have found that in the West in the 21st century, we eat too much food overall and, specifically, too much processed food[5] and sugar.[6]

At the conclusion of one of my lectures, a 40-year-old woman approached me, saying that she worked at a senior level in a high-tech company. She said, "Lately I have experienced a sharp decline in my memory. What could be the reason for this?"

After I was convinced it was not her diet that was the problem, I asked her how she was sleeping. Her answer did not surprise me. "I suffer from severe sleep disorders."

Scientific studies have long found that sleep, sufficient in both quantity and quality, is essential for maintaining our brain hardware.[7] Proper nutrition and good sleep are important components when we want to improve our memory, but they are not enough on their own.

Our emotions play an equally important role, and the explanation for this begins with chemicals. One of the hormones our body produces is cortisol, which has many important jobs. Cortisol regulates blood pressure and the immune system. In situations of stress, trauma and mental distress, our cortisol level increases. Studies have revealed that if we have high levels of cortisol in our body, over time, it causes damage to different areas of the brain, including our memory networks.[8]

It is convenient for us to dismiss these findings, saying, "There is nothing we can do – tension and stress are inevitable parts of our frenetic modern lives." There is a degree of truth to this, but we should keep it in mind and reduce stress as much as we can to help our memory. If you find it difficult to do this on your own, seek help from skilled therapists and counsellors.

So far, we have three components – nutrition, sleep and our emotions – but we need to add a fourth. It is one that is entirely in our own hands: physical activity. There's no need to do gruelling training for a marathon – much less exercise than that is enough. Studies have found that as little as half an hour of walking four to five times a week significantly improves memory.[9] We now have a plausible explanation for this, although there is not unequivocal scientific proof yet.

The explanation goes like this. Our ancient ancestors, who were hunter-gatherers, had to wander great distances to find food, but estimates of how far vary. According to one study, based in part on the examination of ancient skeletons, on average, they roamed 5–6 miles (8–10km) each day.[10] At the other end of the scale, researchers who followed contemporary hunter-gatherers noted that, in their search for food, they walked approximately 18 miles (29km) a day, and they assume our ancestors would have walked similar distances.[11]

Whether it was 5 or 18 miles (9 or 29km), there is little doubt that our ancestors walked a lot, travelling from camp to camp, ranging far and wide to look for food. If they were to survive, they had to remember their surroundings well – sources of water, hunting areas and concentrations of fruit bushes, trees and other food. It is reasonable to assume that, as the brain is one large neural network, a direct connection was created between the motor activity of walking and memory and, with subsequent generations, the connection between the two became stronger.

So walking has a positive effect on the quality of our memory. Of course, genetics also have a great influence on the brain and memory, but as our DNA is an inborn aspect of ourselves that we have no influence over, we will not deal with it here.

No matter how important the hardware is, maintaining it properly and improving it are only half the job. As I mentioned earlier, the brain also includes a software component, and to achieve the maximum and optimal utilization of the brain, we need to take care of this part too. Let's look at how we can do this now.

We already know that memory in itself is not interesting to the brain. Memory is only of interest if it helps to promote our own survival and the possibility that we will ensure the human race survives too. Therefore, if we can convince the brain that a certain message we wish to preserve is essential for our survival, we know that it will be encoded, stored and can be retrieved when needed. All this will only be done in an optimal way if the message reaches the brain in the form of its own internal language. This language relies on five aspects that convey to the brain a message is essential.

Motivation

Motivation is fired up in us when we have reasons and motives to do a certain thing. Anatomically, this happens in a neural network found in the frontal areas of the cortex and subcortex. When we are motivated to remember something – to lock the door when we leave home, for example – the message will reach that neural network and the brain will mobilize according to its internal algorithm to implement the message. It does this because it associates locking the door with our survival.

We can use motivation to play a trick on our brain to help us remember. Consider the following example.

When someone introduces themselves to you, ask yourself why it is important that you remember this person's name. Simply by asking the question – it doesn't matter what the answer is – you will energize the motivation network. The reason for this is simple: by initiating this reaction, the brain receives a message that this information is important for our survival.

There is another way to stimulate motivation, and it is not simply some conclusion that has come to us from brain research. People who work in marketing were way ahead of the researchers in this. If you take a look at some of the promotional materials that tend to clutter your inbox or come through your letterbox, you will notice headlines with the words "New!!! Important!!! Sale!!!" They all have a single goal: to activate your motivation network. To indicate, even if indirectly, that your chances of survival will be higher if you keep looking at the email or leaflet. In short, when we tell ourselves that a certain thing is new or important, we increase its importance in our own eyes, which greatly increases the probability that we will remember it.

Concentration

Concentration involves focusing on what's important. Here, too, the corresponding neural network is located in the frontal area of the cortex, which, when stimulated by receiving the message, helps us to remember it.

When we focus our attention on a certain message, our mind understands, via its internal language, that here is something related to our survival, so it will remember it in the best possible way. To clarify, let's return to my earlier example of locking the door.

When I got home from work, I locked the door behind me. However, later that night, as I was getting ready for bed, I wondered if I had locked the door. Why wasn't I sure about this?

Because I was not consciously focusing on doing this while I was locking the door. Concentration goes beyond simple awareness. As we saw earlier, it corresponds with a defined area in the brain, which must be activated if the memory is to be preserved. Therefore, an effective technique for improving your memory is to concentrate. How do you do it? Very simple. When I lock the door, I tell myself that I have done it. The use of words is not necessary in itself, but the focused awareness that comes with it is what will activate my concentration and, therefore, help me to retrieve my memory later so I know I have locked the door.

Imagination

To activate emotional imagination, I use a scenario that evokes strong emotions. This conveys to my brain, in its internal language, that a certain external stimulus is of great importance to my survival.

My example involves a door again, but this time it's my car door. Immediately after I lock the car, I close my eyes for a few seconds and imagine sitting in the locked car and an armed attacker approaching. They violently try to open the door, but it's locked, so I'm safe. This image creates a clear sense of a threat to my life and connects it to the locked door. Therefore, when I wonder during the day whether I locked the car door, I will see in my mind's eye the assailant who failed in their mission and so I will know with certainty that I did lock the door.

This example teaches us that we can boost our memory if we combine it with emotional imagination. But that is not all – our story has a sequel.

If, on the following day, I want to remember whether I locked the car, and imagine the failed attack again, it will be less effective. This is because when I recall the armed attacker, I will indeed see them, but I might think they're the one from yesterday. There's another trick to solve this problem.

When I close my eyes the next day to imagine something after locking the car, I substitute a gang of friendly gnomes for the attacker, and this time they want to get into the car to valet the upholstery. This creates a completely different emotion from the previous scenario I imagined, but it's just as effective at helping me to remember. When I wonder, "Did I lock the car?" I will see in my mind's eye the smiling faces of the gnomes and remember that I definitely locked the car.

At first glance, this seems strange. After all, those lovable gnomes are no threat to my survival. On the contrary, thinking about them makes me smile, but that's exactly the point. For the brain, the imaginary encounter with the gnomes is still emotionally charged but in a positive way. It represents a moment when surviving is being enjoyed, things are good. The brain wants to capture this enjoyable moment and know how to create similar moments to improve our survival. A combination of any emotion, positive or negative, will improve memory from the standpoint of motives related to survival.

Use a combination of channels

Activating several senses to internalize a certain message will result in it being received at multiple locations in the declarative memory network, creating a robust memory. Let's look at an example of how this might be used in everyday life.

A man – let's call him Oren – introduces himself to us by name. His name will reach the memory network in the brain via the auditory channel. If we shake his hand and concentrate on the strength of the touch of his hand, the name message will also be transmitted through the sensory channel. For the purpose of our example, we will not be satisfied with this to secure Oren's name in our memory, so we might activate our sense of smell, concentrating on this aspect alongside information being conveyed via the other channels throughout the encounter. If we're lucky, Oren will be wearing a pleasant cologne but, equally, we might pick up his natural fragrance. The brain does not care what our noses experience, it is only interested in the message. As the declarative memory network is being bombarded from several channels, the understanding in terms of the brain code will be that Oren is a concept of great importance, so he will be etched firmly in our memory.

Neuroscientist Chris Frith went further in this regard. In his book *Making up the Mind*,[12] he says that he got into the habit of having some sour sweets in his pocket. After meeting a new person, he put a sweet in his mouth while silently muttering the person's name to himself. In this way, he combines the sense of taste with hearing the message, boosting his memory for names.

Using multiple senses together to improve memory is not the only combination of channels available to us. Linking new with old can also be effective. To understand this concept, let's revisit our meeting with Oren. When I heard his name, it had an existing resonance for me as this is the name of one of my nephews, so I reminded myself of this. Because the name Oren is already deep in my declarative memory network, as it is part of my autobiographical memory related to survival, my brain will attach great value to my new acquaintance having the same name and retrieve it easily when I need it again.

False memories

As we have seen, memories are the products of sensory and verbal stimuli that are processed as factual messages and registered in various neural networks. But there is another type of memory: the false memory. Such memories, as their name indicates, are of events that did not happen and so were not created in the usual way. However, if I were to undergo a polygraph test for a false memory, such as a basketball game that I'm sure I watched a few weeks ago, I would pass the test and appear to have told the truth. Yet if someone were to check a video of the game and compare it to my responses during the polygraph, it would become clear to me that there is no way I was at that game.

Until the end of the 20th century, the accepted opinion among brain researchers was that false memories are the result of a brain malfunction. Fascinating experiments conducted by psychologist Elizabeth Loftus proved that this was not a valid explanation.[13]

In one of the experiments, Loftus invited students and their mothers to short interviews. First, she spent a few minutes with the mother and had a completely casual conversation with her that did not refer to her son or daughter at all.

Then Loftus brought the student into the room to join her and said, "I had an interesting conversation with your mother. We talked mainly about you. She told me, among other things, about something that happened to you when you were five years old. You were wandering in the mall and suddenly you got lost . . ." At this point, Loftus stopped and turned to the student, saying, "Maybe you can tell me what exactly happened there . . ."

Even though the event was completely fake, the vast majority of the students (more than 90 per cent) without hesitation gave Loftus a detailed description of getting lost at the mall. A polygraph test would clearly establish that they believed in their hearts the truth of what was a fictitious story. The conclusion drawn from this was that perfectly normal brains can generate false memories.

It confirmed what neuroscientists knew about this phenomenon: the brain is not interested in the past, only the future. Thus, the students' brains must have followed the precedent that if their mother remembers a story, it must be important information. Then, once the brain used concentration and emotion to attach importance to the story, it adopted the false information as part of the neural network of memory. Taking this one step further, in each case, the student's brain would have received the story as being part of their personal history. A piece of information helpful for their survival would be remembered, making it seem real.

A normative mind (one that is generally aligned with the norms of society) will act this way whenever it believes that certain information will help with the survival of its owner. A chapter in "personal history" fits right into this category. Experiments we conducted at the Weizmann Institute unequivocally determined that, in terms of brain function, there is no difference between a real memory and a false memory.[14]

Too much of a good thing

We tend to envy people who have an amazing memory and marvel at their ability to effortlessly recall names, dates and other data as though they were a walking encyclopaedia. People who fall into this category often deserve our compassion as their extraordinary ability often disrupts their lives.

Why is this? The answer lies in the fact that the key word when it comes to everything related to brain function is proportionality. Effective management of the memory networks involves the brain distinguishing the necessary signals from noise in the vast store of long-term memories and bringing to our awareness only the memory relevant to our survival at that moment. An excess of memories, disproportionate to what is useful and needed, burdens the activity of the brain and, as a result, makes it difficult to live your life.

This may be hard to imagine, so think of a young man with memory overload who invites a young woman out on a first date. He firmly believes that he has finally found the partner he's been searching for, but then his excellent memory gets in the way. Memories of previous relationships that ended badly are recalled and derail his feelings. The result? That first date is the last. Even when this young man sits down for a meal, no matter how tasty it is, unpleasant memories of dishes he ate years ago overtake him, so he can't enjoy his food.

Having an excessive number of memories makes it difficult for the brain to fulfil its main mission to help us to survive. That is why our brain invests a lot of energy in proper forgetting. The emphasis here is on the proper balance. An excess of forgetting is as problematic as an excess of remembering. Both create the possibility of failures in our ability to survive.

Forgetting and simulated forgetting

Proper forgetting is achieved by filtering. The first screening of memories happens at the very outset, before even the encoding stage. Our concentration is the mediator, defining for the brain what is relevant and worth encoding. We might be aware of a certain message, but not focused on it, and so it will not be encoded and it will fade away. Also, a message may pass the encoding stage, but its place in the memory network is still not guaranteed. The next filtering step takes place every night, while we are asleep. The brain selects details from the memory network that it deems unnecessary for survival and ejects them.

Proper forgetting is sometimes temporary or, more precisely, simulated, as the following example shows.

When Jackie was at university, 25 years ago, her course included statistics. She has since been an English teacher for many years. One day, her daughter, Kate, asks for help with some homework on statistics, as she remembers her mum mentioning that she had studied it.

"But that was a million years ago," Jackie responds, "I don't remember it anymore."

Although it seems that this is the case, Kate is insistent and so Jackie agrees to help her. After spending a few days going through a relevant textbook, Jackie says that her knowledge of statistics is starting to come back to her.

This is a clear example of simulated forgetting. When Jackie was a student, the knowledge about statistics that she learned was encoded and stored in the correct part of her neural network as a long-term memory. Like all long-term memories, it contained the recipe for recreating that memory. We can think of this as being similar to receiving a double gift – a cake as well as the recipe for it. A memory recipe is preserved for a lifetime and, like a seed before germination, does not require a lot of energy to preserve it until it's needed. The memory is maintained in the brain by the continual supply of nourishment in the form of oxygen, vitamins and proteins that travel in the bloodstream. Jackie had not needed this memory for a long time and it was not necessary for her survival in the present, so full cultivation of the memory would have been a waste of energy. This was directed instead at maintaining other memories that were in constant use for survival purposes. The brain kept the memories of statistics in a state like that of a recipe or dormant seed. The recipe and seed exist but the memories themselves did not exist. When a new need for such memories arises, the brain simply goes back and bakes it from the recipe or grows it from the seed.

Another example of simulated forgetting is when people born in one country relocate to another country at the age of six or seven, and they no longer speak their first language. When asked about this, they will say that they have totally forgotten it all but, often, the opposite can be true. In many cases, a visit to the country of their birth will work wonders. In a few days, they begin to understand what people are saying and soon find that they can converse with the locals.

The process that is happening here with a language is similar to Jackie's experience with her knowledge of statistics. She, too, was sure that she had forgotten it.

Forgetting does not always involve deletion of a memory. Often the brain is satisfied with simply hiding a memory temporarily from our awareness. An example might be helpful here.

A teacher needs to go to her school to prepare the children in her class for their exams. However, that morning her young son wakes up and is unwell. Just in time, she manages to find someone to take care of him.

While driving to the school, she is troubled by thoughts of how her little boy will be while she is gone, but once she stands in front of her class, engrossed in teaching, those thoughts disappear from her mind. Only at the end of the lesson, when the bell rings and the children go, does she remember him, and she quickly calls the friend who stepped in to help babysit to ask how he is.

Remember, the brain exists solely for the purpose of promoting our survival so we can function properly. Temporary forgetting did its job in helping the teacher to do exactly that.

I had the privilege of participating in research at the Weizmann Institute led by Yadin Dodai and Avi Mendelsohn, with Alexander Solomonovich also participating. During the course of this research, it was discovered that an area at the front of the brain is used to suppress memory temporarily. The editors of the journal in which the findings were published found an apt name for this tiny area (it is only a few millimetres across): they called it "memory's gatekeeper".[15]

Do you remember the man we encountered at the beginning of this chapter who was anxious about his memory? Even if you have forgotten him, there is no need to worry. As you will now understand, the one and only purpose of the brain is to help us to survive. From that point of view, having a memory of a person you do not know is of no importance to you, and so forgetting him is proper forgetting.

I hope that if the silver-haired gentleman I mentioned reads this chapter, he will be convinced that it is best to let go of memory tests, that they are not helpful. I have that same hope for you also.

Some more tips for improving your memory

Developing intuition

You can learn to apply many skills using your intuition, to make the subcortical memory and learning processes that operate in your brain work for you. For example, being able to navigate, park your car, recognize untrustworthy people and play music by ear are skills that can all be performed with minimal input from the cortex. So how can intuition be improved to help you do these things?

1 Emphasize to yourself the existing rules for a skill that you want to improve. To navigate to somewhere you want to go, you could have a fixed route from, for example, your home to the pizzeria. To park your car, you could focus on the movements of the car governed by turning the steering wheel and using the pedals. To recognize untrustworthy people, you could look for evidence of the emotion of greed – a characteristic of

such people – which can be seen in their facial expressions. To play music by ear, it is helpful to know that there is a close connection between a certain melody and the notes and chords it contains. The neural network in the subcortex will be able to recognize the existing regularity of the rules of such skills and, subsequently, promote them being used in an intuitive rather than a conscious way.

2 Gain real-life experience of the skills you want to get better at. Ensure you have multiple repetitions of them, accompanied by instant feedback regarding your success or failure when performing those skills. For example, if I drive around looking for the pizzeria without using satnav, even if I make mistakes along the way, this will promote intuitive learning of how to navigate my local area. Also, if I make a lot of stops while doing this, the subcortex will be able to sharpen the rules for this skill and, eventually, I will find the pizzeria, as a result of learning from the successes and failures I have had along the way. Similarly, regarding being able to read people, the more we notice people's facial expressions and interact socially, the more we will be able to understand and learn to use visual information to successfully identify people of dubious character. With playing music by ear, the more we play familiar songs without the sheet music, the more we will be able to play familiar songs intuitively rather than need to have the music there to play them.

We can all improve our intuitive brain skills, and learning will occur naturally as we experience life. This is neuroplasticity (the brain's capacity to change how it responds to stimuli) working in our favour.

Improving the capacity of our memory

According to the brain code, a message received by the brain and perceived to be important to our survival will be a more durable memory than one it deems to be of lesser importance. To convince our brain that the information we want to remember is important for our survival (immediately or in the long term), we need to emphasize its importance. We can do this by employing several brain functions because such a combination is a signal to the brain that the message is of great importance. Here are some ideas to try.

1 **The power of concentration** – Concentrating for a few seconds will increase our awareness, so we can dedicate ourselves to the message we want to remember.

2 **Connecting to emotion** – Create some excitement connected to the message (remember the car-washing gnomes?).

3 **Involving several senses** – If a message is conveyed by more than one sense, this will help to ensure that it is encoded in the memory network. For example, making a point to look at your sunglasses and touch them before putting them down can circumvent you later asking, "Where did I put my glasses?"

4 **Combining a new memory with an old one** – Locating a memory already firmly lodged in your brain that is related to the message you want to remember and linking them can be very effective. So, if you put your sunglasses in a drawer, you could emphasize to yourself that they are in the drawer where you usually keep your keys.

***

Now that the gates to an improved memory have been opened to us, in chapter 3 we will improve our knowledge of how our emotions function so we can label the moments that are worth remembering.

EMOTION REIGNS SUPREME

"I am a rational person . . ."

This declaration is familiar to us all. It seems there are few people who have never said this about themselves. Most of us are convinced that rationality – a word stemming from the Latin *rationalis*, meaning "logical" or "reasonable" – governs our lives. I'm sorry to disappoint you, but brain research disproves this assumption. The findings clearly show that, as it was for our ancestors, emotion rules over us all. Reason is largely used as a fig leaf to explain our actions.[1] The good news is that the flexibility of the brain allows it to control emotion. In other words, we have the ability to transform emotion from being a leading actor into a co-star who listens to the director's suggestions. To achieve this, we must first get to know more about emotion.

Emotion and language

Emotion is a brain function that creates an experience – fear, curiosity, frustration, longing, optimism, pessimism or one of hundreds of others too numerous to mention. Each type of experience is organized as an independent neural network in the brain. This reticular system (consisting of fine networks) is not unique to us humans but has existed for millions of years in all life forms that have a brain. However, our emotional range has become highly developed. There are emotions that seem to be unique to humans, such as gratitude, compassion and altruism.

Although language occupies a central place in our culture, it plays a secondary role to emotion. At best, language is capable of *describing* an emotion, but if the relevant neural network in the brain has not been activated, the emotional experience itself will not exist for the person concerned. For example, a politician may make a public statement that they are confident about the result of an upcoming election, while inside feeling fear and despair about how things will turn out. If the politician's neural network in the area of confidence has not been activated, what they say is only a statement from their lips to the outside world, not what they feel. Based on our experience with politicians, we are likely to shrug our shoulders and utter, "Sure, that's what they all say."

Two vital emotions: pleasure and fear

Pleasure and fear are two basic emotions that all animals are blessed with as a brain experience. This is the case even for those that the evolutionary process has equipped with only a subcortex, such as fish and reptiles.

These emotions have been key throughout the evolution of so many species. They provide the brain with ongoing vital information about everything related to reproduction and survival. Pleasure indicates that the current situation is good, while fear warns of considerable difficulties that could affect the ability to survive and reproduce.

When the areas in the human brain relating to pleasure and fear are activated, this information flows into the fabric of the neural networks, triggering branches that are assigned to a wide spectrum of mental and physical activities.

Pleasure

Eating and sex activate the pleasure zone of the brain. The urge to satisfy desires for them having been created to ensure our survival and that we reproduce. A snake, which does not have a cortex, lacks the ability to think about whether it *should* eat. It only eats because the food gives it pleasure. The same goes for sex: in the absence of a cortex, the concern to ensure the continuation of the species is not a snake's primary reason for seeking a mate.

It often happens that impulses created as a result of pleasure can get out of control. We see this with alcoholism and drug addiction, for example. When a person drinks alcohol or takes drugs, this activates the pleasure zone. The addict's brain networks do not know that alcohol and drugs are dangerous and have terrible side effects. All they register is that these substances give pleasure. In the course of many generations of evolution, a link has been cultivated between pleasure and survival. The brain transmits the message to all its centres not to stop this pursuit of pleasure. This is the source of addiction. The entire brain is involved in organizing things to ensure that the experience of pleasure is repeated. That is why people will do everything possible, to the point of losing their humanity, to obtain their next drink or fix.

While they were going through withdrawal, I had the opportunity to talk to some people addicted to drugs. They told me how far they would go to get drugs – selling all their possessions, stealing money from their parents, begging on street corners and breaking into houses. One of them told me how he climbed a drainpipe to the top floor of a building and crawled through a narrow window into a flat to steal valuables. "Look at me," the frail-looking man told me, "I can't understand how I had the strength to do that."

Considerable resources have been invested in research to achieve a better understanding of addiction and find effective detox methods. Apart from drug addiction, one impetus for this is concern about the enormous damage to the economy and society inherent in gambling – another common addiction. For example, in Israel in 2002, a bank collapse was triggered by an employee embezzling more than $65 million.[2] She told the court that the money was to pay off her brother's huge gambling debts.

Gambling addicts know that they risk losing all their money, but the thrill experienced by the very act of betting and the possibility of financial reward stimulates their pleasure centre and tips the scales in favour of going ahead rather than stopping. Addicts' behaviour is completely different from that of occasional gamblers, who might buy a lottery ticket or visit a casino while on holiday. Gambling addicts will not hesitate to recklessly withdraw more and more money. The bigger the stake, the more pleasure they feel. In contrast, the motivation of occasional gamblers is simply to make some money so, usually, they will not be tempted to take excessive risks. The self-control of occasional gamblers stems from the rational part of their brain being dominant in the situation, and this is missing from the process gambling addicts go through.

We will revisit this later, in chapter 11 on eating habits, and see how proper control of the brain can help people to eat the right foods in healthy amounts. For now, the key point to remember is the enormous importance that the brain places on pleasure because it relates to the motivation the brain has to ensure we survive and reproduce to continue the human species. It is also important to remember the problematic nature of this emotion, which we have touched on in contexts such as drug addiction and compulsive gambling.

Fear

The other basic emotion that steers the daily life of all animals equipped with a brain, including us humans, is fear.

To understand this emotion and the enormous importance it has, we must first introduce ourselves to a structure located deep within our brain called the amygdala. Its name derives from the Greek word for almond, which gives you an idea of its shape. The amygdala is a neural network located in the subcortex. Every external stimulus reaches the brain as an electrical message and moves along two paths. The first path is the fastest, which is the route to the amygdala. This is where the threat is assessed and the correct reaction ensues. The second, slower path leads to the cortex, which is where the cognitive processes take place to decipher each message.

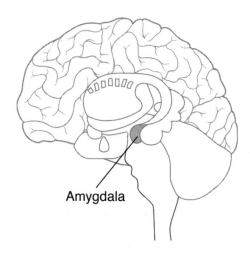

Amygdala

As a way to picture this process in action, I will introduce you to my neighbour, Moshe, who tends his small garden. Imagine Moshe is doing some weeding and hears a rustling noise. When he looks around, he sees a long, black shape beneath the bushes. For a fraction of a second, he thinks that there is a snake in the garden. His heart rate surges to prepare him to escape. But, a second later, Moshe recognizes that the black shape is nothing but an irrigation pipe, so he begins to relax and smiles at his panicked reaction.

What occurred here is that, in the first fraction of a second, the fastest information transfer pathway was activated, causing the amygdala to sound an alarm that rang in all the subcortex areas to which it is connected. They are the motor, sensory, cognitive channels and those that have control over the body's organs. They automatically primed Moshe to defend himself against a snake. Meanwhile, the same information reached the cortex in the next second and was processed, an accurate identification was made and a new message was relayed – "Relax, it's only a pipe!"

Here's another story to illustrate the way in which the brain treats a perceived threat.

While driving home late at night, I had a feeling that the keys to my flat were not where I put them in the car. Clearly, if the keys were not in the car, I would have a major problem, so my body reacted accordingly. The amygdala assessed the disappearance of the keys to be a threat and I felt alarmed. When I got home, I stopped the car and started looking for my keys. Then, all of a sudden, a feeling of relief swept over me.

"Strange", I said to myself, because nothing had happened to allay my fear and I still hadn't found the keys. Then I saw them . . .

Let's analyse why I felt relieved *before* I found my keys. It is because the message from my sense of sight reached the subcortex before I was even aware of it and silenced the alarm bell that had been sounded by the amygdala in response to my thought that I did not have my keys. The amygdala's ability to detect threats is, first and foremost, genetic and similar to what happens in animals without a cortex. Certain stimuli (including smells, intense heat, intense cold, loud voices and fast movements) will activate the amygdala even in babies, whose cortex is not yet mature. As the cortex develops, the brain assigns meaning to other basic stimuli, such as explosions, a passing car, a speeding motorcycle, burning and the like.

One of the most important areas of the brain activated by the amygdala is the hypothalamus. The hypothalamus is a complex neural network located in the subcortex and is activated, among other things, when the amygdala transmits a message that there is a threat. The hypothalamus has two branches: one regulates behaviour in the face of threats (the behavioural branch), which makes it possible for us to respond appropriately to such a situation. The other is in charge of the secretion of hormones (the hormonal branch), which signals to the body that it is facing danger and must prepare for the worst to ensure the optimal management of an emergency.

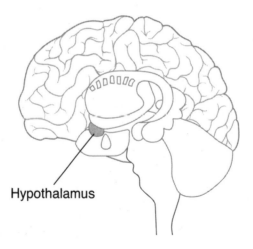

Hypothalamus

The three F strategies

The behavioural branch can command the body to react in one of three ways. These reactions are known as the three F strategies, which are:

- **Fight** what threatens you
- **Flight** – or flee – from danger
- **Freeze** in place

Depending on each individual's genetics and life circumstances, one of these strategies will be dominant. Thus, when the hypothalamus is activated, some will prefer to fight, some will be more inclined to flee and others will freeze. Unlike in other animals, the developed human brain not only determines which F strategy to use but also considers the amount of force to employ.

To understand how this works, let's examine the behaviour of two people, both of whom are in a situation that requires a fighting strategy. One immediately starts throwing punches, while the other is content with making a cynical remark. Both employ the same strategy but with a different amount of force. It is important to note that the roles of genetics and life circumstances are not limited to choosing the strategy; they also determine the amount of force used.

Adapting the strength of our response to a particular threat requires mental flexibility. Sometimes we need to be very forceful to counter a threat; in other cases, an appropriate verbal response is enough. If someone launches a physical attack on you, the practical response is either fight or flight. Nevertheless, if your boss, on whom your livelihood depends, raises their voice, it is better to respond by freezing until their anger fades rather than employing a fight or flight strategy. However, people with a rigid personality often adopt the same strategy whatever the threat situation – always fight with a high level of intensity or always flee or always freeze. The greater someone's emotional intelligence, the more accurately they will gauge how to adapt a strategy to deal with a given situation.

The choice made between the three strategies is automatic, and so is the level of the reaction. For example, a member of staff working at the checkout in a supermarket glances at the bag of a customer standing in front of her. The customer feels threatened, thinking that the member of staff suspects they stole something. The strategy dominating the customer is to fight. So, getting bolshy, the customer starts shouting at the member of staff.

If, after the fact, the customer realizes that they overreacted and should adopt a more balanced way to behave in such circumstances, the brain's flexibility will allow this to happen. Merely wanting to do this will not make it happen. Concentrating on the idea that they must moderate their reaction, and being deeply motivated to do so, will make it clear to the brain that this is important. Only then can they make such a change to their behaviour.

Cortisol

The activation of each of the three F strategies requires an increase in the supply of hormones. This job is assigned to the other branch of the hypothalamus, the hormonal one, which, as you would expect, regulates the release of the relevant hormones from the glands into the bloodstream.

One of the hormones regulated by the hypothalamus following signals received from the amygdala is cortisol. This chemical is needed for the optimal activation of muscles as part of the strategy chosen to deal with a threat. The problem is that persistently high levels of cortisol over time damage the brain and other systems in the body. As long as the threat is not overwhelming and frequent, there is no cause for concern. It is when the threatening situations pile up and are not properly regulated that cortisol switches from being a friend to a bitter enemy.

These situations – of feeling dangerously threatened for long periods of time – result from the fact that the mechanism used in the brain to activate the amygdala and the hypothalamus has not changed for millions of years. We still come into this world equipped to handle the same patterns of activity that were important to our ancient ancestors.

However, being set up in this way no longer fits the environment we live in today, nor our lifestyles.

To get some idea of how this pattern came to be established, imagine that you are with one of your early ancestors. While you are hunting with him across the savannah for the family's next meal, he notices a wild boar coming dangerously near. The neural network receiving this vital message from his eyes convey it to his amygdala, which causes him to spring into action without any hesitation to deal with this threat. As a result, he is in optimal flight mode. The cortisol levels in his blood will be very high to allow the muscles to work vigorously to evade the wild boar. Fortunately, the hunter does not become the hunted, and when the danger passes, the amygdala returns to a state of calm. His cortisol levels drop back to normal and, as the threat was short-lived, the elevation in this hormone has not caused any damage to his body or brain.

While this operating pattern for the brain has remained the default ever since those early days of human existence, to protect us, the threats we have to deal with in the 21st century are of a completely different nature. To face down a modern threat, we will now accompany one of the hunter's descendants – a hard-working programmer called Luca.

One day, Luca's team leader convenes a staff meeting and announces that she's heard rumours of members of staff being laid off. "I have no idea when this might happen," she emphasizes, "nor do I know how many people are involved, but I thought I should tell you so that you can prepare for such a possibility."

As the brain code does not recognize intermediate states, this news is perceived by Luca as a high-level existential threat. This causes his hypothalamus to kick in, resulting in the secretion of cortisol and other stress hormones. But, unlike the hunter, this flow of cortisol into the bloodstream will not be a short burst. The threat will not be over quickly. In fact, it will continue until the unsettled state at Luca's workplace is sorted out. Even if this finally happens and he still has a job, new threats will appear, such as a looming deadline for submitting a proposal, and less serious situations – being stuck in traffic on the way to work, for example.

By comparing the distant past to our environment today, we can see how modern lifestyles and culture have greatly expanded the concepts of what is a threat and survival. There can be multiple events in our lives that we perceive as threatening, so many people store much more cortisol in their bodies than is necessary. All this cortisol flooding our blood is a ticking time bomb that can have devastating consequences for the body. These range from heart disease, diabetes or brain dysfunction to cancer, stemming from disruptions to the immune system. Is this the fate that 21st-century human beings are destined for? Do we have no choice but to accept it?

Strategies for beating fear

The good news is that, in the age of the flexible brain, things can be different. We can train our amygdala to function in appropriate intermediate states – that is, take control of how our emotions function. Let's return to Luca the programmer to see how this works.

The team leader's statement created a problematic situation, but there's no immediate threat to Luca's survival – it seems to be far off in the future. It is likely that, after the first shock, he will tell himself (and his friends will tell him too) that the situation isn't so terrible. The job losses are not definite and, even if he is fired, it is likely that his programming skills will help him to find a new job fairly quickly, maybe even one that pays better. If he is let go and it takes time for him to find a new job, he will have a few months of unemployment benefit and he could use the time to bake cakes. Baking is a hobby that he's not been able to enjoy recently as he's been too busy at work.

Luca repeats these thoughts to himself. He tends to believe it, but words by themselves do not lead to changes in brain activity. His amygdala, still in threat mode, causes the hypothalamus to order the adrenal gland to secrete large amounts of cortisol. So what can be done to regulate the amygdala to respond to the *real* level of threat and lower the levels of cortisol produced?

Connect to the threat

The first thing Luca must do to control a threat is to become familiar with it. It is not enough for us simply to try to understand on a rational level why we feel threatened; we must connect to the threat. We must experience it and delve into it. This can be done by focusing on other threatening events that we have experienced in our lives. When doing this, it is important to choose events that led to a positive outcome so that the experience is constructive.

This encourages the creation of two-way neural pathways between the awareness areas in the cortex and the amygdala in the subcortex. When our attention is directed to emotional experiences, they're perceived by the brain as promoting our survival. It will reinforce these pathways that enable a direct flow of information from the amygdala to the cortex. After paving the way between the amygdala and the awareness areas in the frontal cortex, information can then flow in the opposite direction too. Information aimed at regulating the amygdala will flow from the cortex to the subcortex, and its effect will be reflected not only in a more balanced activity of the amygdala but also in a more measured response to the intensity of the threat than the high alert observed so far.

The fear regulation switch

Another action that can be taken is to use a technique called the fear regulation switch. The threat of dismissal struck Luca powerfully. If he examines what is an unexpected state of affairs, he will likely conclude that the correct level for this threat on a scale from one to ten is, at most three.

A simple and effective way to do this is to imagine a dial that regulates the intensity of a threat, and it is currently set at ten. Using the power of imagination, we can slowly turn the dial and reduce the power of the threat until it is at an appropriate level. Neutralizing the amygdala will not stave off the threat of dismissal, but reducing its intensity will allow Luca to get his life back to normal and lower the level of cortisol in his body to less harmful levels.

Motivation and meaning in the control of emotion

Pleasure and fear are at opposite ends of the emotional spectrum. When pleasure is active, survival seems guaranteed and life is sweet. When fear kicks in, we sense the danger of extinction and feel stressed and tense.

In everyday life, we're continually doing things, often simultaneously, that may provoke us to react with different levels of these two emotions. At any given moment we can find ourselves at certain points on this spectrum, between pleasure and fear. The brain strives to get us as close as possible to the pleasure end, by making us feel like we are scoring points every time we move in that direction. It does this because it is working to increase our chances of survival.

Imagine that you are looking at a cup of tea on my desk. I'm engrossed in my writing and, as I haven't taken a sip for a while, my current survival score is reducing. Distracted, my hand reaches out for the cup of tea and, even though I was not aware of it at all, my score starts to improve, as I am increasing my chances of pleasure and survival when I take a drink. My brain perceives that my survival score has improved, so it activates the motivation area of my brain to persuade me to continue along this path. As my hand gets closer to the cup, the score increases again. When I drink, the emotion of pleasure is at its peak. The motivation to grab the cup of tea and bring it to my mouth was not something that I was consciously aware of. It was the score my brain gave me, without me being aware of it, that created this motivation to drink.

Most of our daily activities are performed without us being fully aware of them. Neuroscientist Chris Frith describes an experiment in his book *Making up the Mind*[3] that was carried out in the Department of Psychology at the University of Cambridge that proves this clearly.

A certain professor, unbeknown to him, was a guinea pig for the experiment. He would stand centre stage during lectures and never step left or right. One day, he was asked to delay his arrival by ten minutes so that members of staff at the university could see to some administrative matters. The organizers of the experiment used this time before the professor arrived to brief the students attending the lecture. Those sitting on the left-hand side of the hall were asked to nod their heads and return the professor's gaze whenever he turned his attention to them. Some of the students sitting on the right-hand side of the hall were told to yawn instead when the professor looked their way, while others were instructed to look at their watches when he looked at them.

The professor duly arrived and the lecture got underway without him suspecting anything. At the end, the experimenters saw that he was way over on the left side of the stage. When asked if he felt anything was strange, he replied, "Absolutely not." When he was told what had happened, he responded, "No way, I wasn't aware of that."

What had happened? Similar to my experience with the cup of tea, it was the subcortex alone that managed the professor's behaviour. The high score on the pleasure scale that the brain awarded in response to the positive reactions of the students sitting to the left of the stage motivated the professor to move toward them. However, he did this without being aware that he was reacting in this way. That's because the conscious mind – the cortex – was not involved in the decision to move in that direction.

The evolutionary process has not changed the way the brain operates since the time of our earliest ancestors, because the pursuit of pleasure without conscious awareness dominates even in creatures that do not have a cortex. However, unlike them, the ability to think allows us, the owners of a developed cortex, to motivate ourselves with awareness. If I promise my students that the one who scores highest in the end-of-term exam will win a prize, their overall grades will be much higher than usual. The awareness of the pleasure inherent in winning a prize will stimulate the part of the brain responsible for motivation and the students will study harder. Salary increases and bonuses at work have the same effect, with employees knuckling down to their tasks when aware that additional income will lead to increased pleasure.

The understanding that increased motivation stemming from meaningful creativity results in better levels of performance by employees has gained a lot of traction. Employees who attach meaning to their work feel they belong and are valued. This feeling activates the promotion of long-term survival in the cortex (in contrast to the reaction in the subcortex, which is only concerned with immediate survival). These employees will not only be efficient but also creative, show initiative and enjoy a contented life. Many companies have applied these findings by giving employees autonomy and meaningful challenges. The results have a positive effect on the balance sheets and speak for themselves.

I once had the opportunity to talk with a police sergeant who patrols a high-crime area. When he told me about his low salary, the long hours he worked and the inherent dangers involved in his job, I couldn't help but ask him if he had considered leaving policing for the private security business.

"There's no way I'm leaving the police," he replied, without thinking twice. "Imagine what will happen here if we all leave one day. One thing's certain, there'd be plenty of work for my friends in the force, the paramedics and the undertakers."

Here is proof of the power of meaning to motivate people. In the sergeant's mind, his work is connected with the long-term survival of the local people, and this motivates him and plays a key role in his decision to stay in the police force.

Another aspect of meaning can be learned from observing creatures that do not have a cortex. Motivation for these animals is based solely on the pursuit of pleasure and the need to escape from threats, and this is reflected in their behaviour. An instructive example is fish that prey on some of the fry they spawn. This sounds terrible to us but, for them, eating the fry spikes the fishes' subcortical pleasure score, and enough offspring survive to ensure the survival of the species.

This does not happen among creatures that have a cortex. They are motivated by a feeling of purpose to care for offspring until they are no longer helpless. Their offspring are not simply organic matter to be eaten, but their own kith and kin to be nurtured and cared for. The same is true of us humans. In addition, we have language, which allows us to articulate the meaningfulness we feel and try to explain the logic of it. For example, new parents caring for their baby day and night will find it tiring, but also be motivated to do it because they know how important it is for their baby to be fed, changed and engaged with for its physical and mental development. Their motivation is not only to keep their baby alive but also their love and sense of purpose, which strengthen their resolve in those difficult times with a new baby. Despite the sleep deprivation, they can still enjoy good health and feelings of wellbeing.

Brain research has proved that people who find meaning in their lives and actions develop a sort of mental immunity that protects them from negative stimuli in everyday life.[4] Negative stimuli can take the form of insults, misunderstandings, quarrels and the like. If all these benefits were not enough, a life lived with meaning increases our chances of attaining happiness.

Happiness is an elusive concept that can be interpreted in different ways. Philosopher Aristotle understood more than 2,000 years ago that happiness stems from a combination of purpose and pleasure. For example, when my partner told me that our daughter had said "Daddy" for the first time, this news activated the emotion of pleasure in me and, because the deep meaning I attribute to my daughter already exists, a combination was created between the two and the feeling of happiness burst forth. Even such joy is short-lived because the brain quickly adapts to the feeling of happiness and it fades away but there will be other such experiences to come.

Some tips for keeping pleasure and fear in balance

The motivation triangle

As far as the brain is concerned, purpose is expressed as activity in the frontal area of the cortex. To be motivated by the purpose of a particular goal, we must first activate the relevant area of the cortex. One method that you can use to do this is based on scientific research and is what I shall call a motivation triangle.[5]

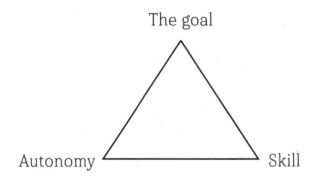

The goal

This is the first element of the motivation triangle, which here can be not snacking after 9pm. To make this goal easier to achieve, we could limit ourselves to keeping to it on weekdays initially. If we use our power of concentration and imagine in detail how those five evenings will look, this will make it clear to the brain that this idea is important to us and so it needs to apply it.

Autonomy

The second element of the motivation triangle, autonomy, means that we act with complete independence and exclusively control the implementation of the decisions we make. Acknowledging this will strengthen our motivation to stick to our goal. After the first five evenings of managing not to snack after 9pm, this will boost our confidence that we can master this skill.

Skill

This is the final element of the motivation triangle. Successfully completing the first round of five evenings will give us the determination needed for a second cycle of five evenings. The more we repeat this practice, the more we will strengthen the connection in the brain of the idea of not snacking in the evening to the area in the brain relating to purpose. What initially was a challenge will become easier. If we should succumb to occasional temptation, there is no need to panic – it happens. We simply have to stick to the plan again and when the deep meaning of the goal has been created, the brain will implement the idea automatically, without the need for us to trick it into doing so.

This method can work successfully in all areas of life. One of its greatest proponents was Viktor Frankl, an Austrian Jewish psychiatrist, psychotherapist and neurologist from Vienna who survived four concentration camps, including Auschwitz. In his famous book, *Man's Search for Meaning*,[6] Frankl asserts that the people who survived the camps were those who developed deep activity in the neural networks associated with meaning. Detailed planning of how they would live their lives after liberation strengthened their motivation to survive far beyond simply the natural desire to stay alive. What kept Frankl alive was his desire to rewrite and publish his book, which was lost at Auschwitz.

***

When the brain's activity is linked to meaning in the cerebral cortex, this also induces optimal brain activity in the subcortex.

In chapter 4, we will learn, among other things, how to activate positive emotional activity in the brain.

CHAPTER 4

THE COLOURS OF EMOTION

Thanks to evolution, the sum total of our human emotions is infinitely greater than that of all other animals. We are blessed with unique emotions such as gratitude, compassion and altruism. Brain researchers looking at the activity of the neural networks in the subcortex have discovered hundreds of emotions that are either positive or negative but we are going to limit ourselves to a small sample. Positive emotions include love, affection, optimism, mercy, caring, forgiveness, longing, loyalty, gratitude, determination, benevolence, joy and curiosity. Negative emotions include hatred, sadness, pessimism, embarrassment, loss, envy, frustration, disappointment, boredom, loneliness, loathing, alienation and defeat.

Emotions can grip us, but it would be incorrect to say that we are always dominated by one particular emotion. The situation is more complicated than that. At any given moment we are influenced by a mix of different strengths of both positive and negative emotions. They combine to create our mood. Each emotion is anchored in a unique neural network, a kind of software with the ability to operate independently. We have no problem simultaneously experiencing different emotions, even if they seem contradictory. For example, as I am writing these lines, I feel both optimistic that this book will interest my readers *and* apprehensive that I might not be expressing myself properly. I'm also full of joy because I've completed several chapters *and* I'm worried that I will not meet the schedule I set for myself. So I feel satisfied with my mastery of the material but also frustrated because time is flying by and there's still a lot to write. As well as joyful knowing that, in a few hours, I can spend some quality time with my young daughter, I'm experiencing feelings of regret because something came up at the last minute and I couldn't keep a promise to meet my mother yesterday. In addition to all these, there are feelings that I am not even aware of.

Having such a range of emotions is the result of evolutionary processes that have formed the brain over millions of years to ensure that human beings survive and reproduce. Every emotion, both positive and negative, is important. So what possible benefit can we derive from negative emotions. The Talmud (the principal source of Jewish religious law) provides a convincing reason. Although envy is unequivocally held to be a negative emotion, Talmudic sages wrote, "Jealousy of scribes breeds wisdom." Envy can allow us to develop in a positive way, but this is conditional on achieving balance. We need the correct intensity of envy for this to happen. The same is true for anger, which, despite being defined as a negative emotion, may be good for us at times. Studies have found that when we are moderately angry – that is, when this emotion is regulated properly – our thoughts are more focused, we express ourselves more clearly and we are more persuasive to those around us than when we are not angry.[1] Moreover, anger may deter those who provoke it in us, so avoiding things escalating into a physical confrontation. It can also help give us sufficient strength to deal with the person making us angry.

Regulating our feelings, so that they are set at the correct emotional temperature, is also important when dealing with positive emotions. Picture in your mind an intensely compassionate person. If they come across an injured cat lying in the street, the overwhelming wave of compassion that will wash over them on seeing the cat in such a state will not allow them to act as normal. Such an overabundance of compassion would prevent this person from being able to act appropriately in professions such as medicine, psychology, education or social work. Similarly, although it would seem to be a positive thing, people who are always brimming with optimism or supreme joy generally have trouble behaving appropriately.

The key word when it comes to emotions is balance. We need to be able to respond with the correct emotion, at the right moment and apply the right dose for the situation we are in. To help with this, it's a good idea to wake up in the morning focusing on a positive range of emotions. Later, if we find ourselves in a situation that requires a negative emotion, it's good to apply it, but remember to use the right amount for the occasion!

All negative and positive emotions can be said to be derivatives of the two basic core emotions – fear and pleasure – and have similar effects on the amygdala. Negative emotions such as frustration and jealousy cause the amygdala to react in such a way that we perceive the environment as threatening, with positive emotions such as joy and affection having no effect on changing that view.

The extent to which we can think rationally and perceive reality correctly will be the result of the feelings we experience at a given moment. Let me illustrate this by exploring what happens when two people, who are total strangers, are sitting at separate tables in a restaurant. One is caught up in a mixture of jealousy, anger and frustration, while the other feels optimism, joy and compassion. Both have to wait a long time for the busy server to come over and take their orders. Let's take a peek into the minds of these two people and examine their thinking in this context.

The diner with the positive mix of emotions tells himself something like, "The poor server is working so hard. I was lucky not to have to work in a restaurant when I was a student. It's OK, I'll just wait my turn."

Meanwhile, the diner with the negative mix of feelings, grumbles to herself, "Why does this always happen to me . . . I sit, I wait and am ignored. The server walks right by me but doesn't even see me! I've had enough of this, I'm fed up."

Even though these two people are having exactly the same experience, the reality is the same for both of them, but they each interpret it in completely different ways. Not only that but each person is convinced their perception of this reality is the correct one.

The diners' perceptions are different because they have different combinations of emotions, and those combinations dictate how rational and balanced their views of the reality are. The variations between the two mixtures of emotions they are feeling have their origins in the mental structures of each person. The first diner's default state is positive so, if there is no special reason for negative emotions, he will experience a medley of positive ones. However, for the second diner, negative emotions are her default and so an extraordinary external stimulus is needed for positive emotions to come to the surface. This difference in default states also dictates their quality of life, cortisol levels, heart rate and blood pressure.

This situation shows the extent to which emotions can colour our perception of reality and have definite effects on our quality of life. Fortunately, they are not beyond our control. Thanks to the marvellous flexibility of the brain, we can use reason to influence our feelings, and so paint our emotional life with brighter colours than the second diner in our story. The intention should not be to shake off negative emotions completely, but to arrive at a situation in which positive feelings are our default, unless there's a genuine reason for negativity, of course.

How to improve your ability to regulate your emotions

To help you regulate your emotions, here is a practical way to strengthen your awareness of your emotions, which will help you control them in a beneficial way. The exercise I am about to describe lasts for 40 consecutive days.

An exercise to enhance your awareness and control of your emotions

First, make a list of 20 types of emotions – 10 positive and 10 negative. Your positive list might include forgiveness, admiration, determination, passion and elation. Examples for the negative list could be resentment, envy or aggression. If you have trouble listing that many emotions, you can search online for positive and negative emotions and you won't be short of options!

There is no need to divide the feelings into the two categories of positive and negative. Nor should you think of them as good and bad feelings. Just as both types are equally important in our daily lives, both are equally important in this exercise.

Once you have your list, you can move on to a new page to start the first day of the exercise, which takes place over the next 20 days. Each day, spend around five minutes thinking about one of the emotions on your list. I recommend that you do this while you're alone, are feeling emotionally at ease and when your attention will not be diverted by some other task, such as driving.

During those five minutes, recall events from your past that are related to the emotion you have chosen. For example, if the first emotion on your list is optimism, let your thoughts return to past events during which you felt optimistic. This does not necessarily mean that the events were in and of themselves optimistic, but they were ones during which optimism was the emotional experience you had.

I recommend that you recall at least three events, and it is advisable to keep notes, writing down the events.

The next day, select another emotion from your list. For example, if the emotion is hatred, devote five minutes to events when you experienced it and proceed as you did on the first day. The next day, continue in the same way, and so on, with emotion after emotion, until you have done this for 20 days and considered all the emotions on your list.

Next, we move on to the second half of the exercise, which requires a little more effort than the first half. Are you ready?

From now on, it is not enough for you to merely remember events when you felt a certain emotion. Instead, you must turn your attention to the emotions themselves, as you experienced them during each of those events.

As we are now on Day 21, let's go back to the first emotion on your list, using optimism as an example, as before.

Close your eyes and recall from your memory the events you wrote down for optimism on the first day of the exercise. Remember the people, surroundings, sounds, sights, smells (if any) and any other relevant details.

By reliving these moments in detail after the previous 20 days of dealing with the emotional world, there's a good chance that you will be able to activate the areas of higher emotions and feel the optimism you experienced back then. I don't want to mislead you, this task is not easy. It requires patience, perseverance and, above all, a great deal of concentration.

Continue in the same way for the remaining 19 days, completing this process for all the events relating to your list of emotions. The more committed you are to this, the easier it will be for you to experience the emotions you associate with the events.

Using what you have learned

This exercise improves the connections in your brain between the areas responsible for awareness and emotions. As a result, not only will your awareness of your emotions be strengthened but your control of them will also be enhanced, and you will be able to call up the correct emotion, with the necessary level of intensity, in any situation. This is because one way to convert a negative feeling such as anxiety into a positive one of optimism is to recall an event from the past in which you experienced optimism. For example, if you fear losing your job, you will find you are able to convert the anxiety that is the natural default feeling for such a situation into an optimistic one instead by anticipating the possibility of moving on to a new, better opportunity. Similarly, in a tense situation involving a family member, the unpleasantness you feel can be transformed into the positivity of seeing it as an opportunity for an open conversation that might bring you closer to them.

From this, it is clear that to make a change to your behaviour, you first need to make a transition from one emotion to another. To do this for our example of the tense situation with a family member, you would concentrate on moments in the past when you felt really close to them. Moments when you trusted them, such as when you offered to help each other and this help was gladly received. In such a case, your motivation of wanting things to be better between you can work to convert the negative emotion into a positive one.

However, again, it is important not to think of negative emotions as terrible monsters that we must suppress or get rid of as quickly as possible. Every emotion, the positive and the negative ones, can benefit us. Therefore, we must allow ourselves to be with each emotion, get to know it, test it and ask ourselves if it suits us in the particular situation in which we find ourselves. Moreover, if we look closely at the reasons for the negative emotion, we can often find ways to use or change the situation that created it and turn it into an advantage. As the saying goes, when life gives you lemons, make lemonade!

Some more ways to improve your ability to regulate your emotions

Now we understand that, in terms of the brain code, emotion rules supreme, let's learn how we can control the height of this flame and accurately regulate our thoughts and behaviour without boiling over. To acquire this skill, we need to invest some time. This is because in our modern world we have neglected the emotional dimension of brain activity, and so to practise regulating our emotions, we need to master a new language.

1 **Increase awareness of your emotions** – To perfect your ability to initiate the brain activity that promotes control of your emotions, first focus your attention on your emotional channel. During the next 20 days, devote a whole day to each of the following types of emotion, remembering the life situations in which you experienced them:

- curiosity
- worry
- courage
- hostility
- longing
- jealousy
- happiness
- missing someone
- satisfaction
- loneliness
- determination
- anxiety
- enthusiasm
- sadness
- serenity
- shame

- reconciliation
- being insulted

- admiration
- freedom

2 **Summon up emotional experiences** – For the next 20 days, dive deeper into your emotional language by dedicating a day to each emotion listed in Step 1 again. Focus on one emotion a day, together with the experience of that emotion you recalled. As you explore the details of your memory, you will feel the emotion emerge. To get the most from this, concentrate on the event – what did you see, what did you hear? Bring to mind the smells, your senses of touch, taste and movement. Allow yourself to inhabit the feeling of each emotion completely.

3 **Regulate the intensity of your emotions** – Having reached this stage of the exercise, the emotional world will now be familiar to your mind and so you can begin to practise regulating the intensity of these emotions. To do this, concentrate on the idea that each emotion can be experienced on a scale from one to ten (one being very low intensity and ten the highest). For the next 20 days, dedicate each day to experiencing each emotion listed in Step 1 at different levels of intensity during that day. For example, on the first day, which will be dedicated to curiosity, begin by recalling a life event in which you experienced curiosity but at a very low level. Later that day, revisit a life event in which the intensity was high – feeling curiosity at its most intense level.

Once you have worked your way to the end of this exercise, you will have reached an advanced stage in training yourself to regulate your emotions. You have been acquiring the ability to transition between emotional intensities. Once your mind has learned this skill, you will be able to regulate the intensity of a particular emotion depending on the situation when you need to. For example, if you were to feel insulted and feel this too intensely, you would be able to regulate that feeling by focusing on a different event in which you felt insulted but to a lesser degree. If your learning progresses, your brain will be able to perform this task of emotional regulation automatically.

***

Remember that the world of the emotions only dominates the world of reason at the level of being the default for the brain. We all possess the tools we need to use reason to control our emotions, and so we can improve our physical and mental health.

CHAPTER 5
STRUGGLES OF THE MIND

"How are you?"

"OK, thanks. And you?"

"Yeah, thank you . . ."

This kind of exchange is familiar to all. For the most part, these are nothing more than polite enquiries in line with accepted social codes. But when a neuroscientist asks someone "What is your mood at the moment?" – they attach great importance to how the person responds. This is because the prevailing understanding among brain researchers is that mood reflects a person's perception of the world at any given time, which is dictated by their motor and cognitive abilities (memory, thinking, concentration and so on) and overall state of health.

Mood and the spectrum of emotions

To understand how this works, remember the simple fact that the brain is one large neural network, in which every function has the potential to be connected to all the other functions. The original evolutionary rationale for the connections between brain functions was what they contributed to our survival. From the moment we are born, there are already connections in our brain between different functions that have proved useful during the long history of humankind. Consider, among other things, the connections between emotional and cognitive areas, between emotional and motor areas, as well as from motor areas to cognition, as we saw in chapter 2 on memory.

Mood is a brain function that is related to our emotional world. In chapter 4, emotions were divided into two groups – positive and negative, pleasure and fear. For the purpose of understanding mood, we will now divide emotions differently. This does not contradict or negate the division used in chapter 4, as will become clear.

In everything related to moods, the most useful distinction we can make is between feelings that create impulses that increase brain activity and those that do not cause the brain to react, leaving us in a passive state. Examples of emotions that lead to activity in the brain are joy and envy. We discern that joy is positive and envy is negative, but dividing them into positive and negative emotions is irrelevant to mood. It is more helpful in this case to place both in the same category of increasing brain activity. Emotions such as satisfaction and sadness do not have this effect, so it is best to place them in the category of moods that leave us in a passive state.

Because the brain's one and only goal is to take care of our survival, at every moment it will choose for us the best mood to achieve this goal. In other words, if the brain "thinks" that right now we must be actively striving for our survival, it will choose a mood from the end of the mood spectrum that will increase brain activity. However, if our survival seems to require a slowing down of brain activity, the mood chosen will be from the passive end of the range.

In our daily lives our moods will shift in this way depending on the circumstances. For example, the appropriate mood for those in mourning is one at the passive, quiet end of the range, because the resulting slowing of brain activity will allow them to focus on processing their grief, which will promote their survival. Nonetheless, if someone has received a promotion at work, a mood from the end of the spectrum that increases brain activity will engender a burst of enthusiasm, which will help lead them to success in the new role. Despite seeming extreme, the moods in these two polar opposite examples are considered normal. In everyday life, a mood we consider to be normal moves to and fro in the middle of the range according to circumstances, and rarely strays from this middle ground.

Although I have used the term "normal" here, it is society that determines the degree of normality associated with different moods and what this is will vary from culture to culture around the world. For example, if extroverted and highly emotional behaviour is standard in a certain society, a normal state of mind will be close to the end of the spectrum that increases brain activity. It follows that people in such a society will generally be happier than those in some others, but the opposite moods will be more pronounced too. They will become angrier, more frustrated, more jealous and more anxious than people in less extrovert cultures. Similarly, in a society where the culture is to follow a more passive way of life, the general mood will be closer to that end of the spectrum, its people tending to be calm, restrained, quiet, relaxed and, perhaps, even sad. However, if someone's dominant mood, wherever they live, is at either one end of the spectrum or the other, and this carries on for longer than is considered appropriate, this would be diagnosed as a mood disorder.

Mood disorders that decrease brain activity

Some mood disorders manifest as depression, causing a problematic slowing of brain activity. Statistically, around 15 per cent of people in the West experience an episode of depression at least once in their lifetime.

Depression does not simply appear for no reason. It is triggered by a life event, and the severity of that event is directly related to the severity of the resulting depression and its duration. For example, if your bank informs you of a debt of a large sum of money that you were not aware of, your mood will immediately swing toward depression. Under such circumstances, being in a low mood for a reasonable period of time – a few hours or perhaps a few days – would be considered normal in relation to Western cultural norms. But if the low mood were to continue for longer than that, it would be thought problematic. Indeed, a depressed mood that persists longer than two weeks is one of the criteria for a diagnosis of clinical depression.

Clinical depression

At its worst, clinical depression completely paralyses emotions and so also numbs any urge to be active. Because of this, people who suffer from severe clinical depression are completely passive. They lie in bed most of the day, do not communicate, think slowly and have difficulty remembering. Even simple actions, such as brushing their teeth, become unbearably difficult. Their response to the question "How do you feel?" will likely be a silent shrug or a muttered, "I don't know." Because of this emotional paralysis, they are not even sad, merely indifferent.

Clinical depression progresses along a continuum, with the description I have given being of those at the extreme end of the spectrum. At more moderate points along it, people will be in a poor mood, with feelings of helplessness, hopelessness, a lack of self-esteem and unable to enjoy things that used to give them pleasure. Negative feelings multiply. Because these emotions still have the potential to drive activity, people in the mild stages of clinical depression are still somewhat active, but the steep decline in their brain functions causes a slowing down in motor, sensory and cognitive ability. It is not uncommon for people in late middle age and suffering from clinical depression to be wrongly diagnosed as having Alzheimer's disease because symptoms such as impairments in memory, thinking and concentration are common to both conditions.

The fear that someone suffering from deep clinical depression will attempt suicide increases at the very moment when their condition seems to be improving a little. This is because, when a person is in the depths of depression, they lack energy to such an extent that they are simply unable to think of ways to end their life. When their mood improves slightly, but the suffering still persists, they then have sufficient energy to contemplate suicide and so may carry it out. Therefore, if it seems to you that someone close to you is depressed, it is appropriate to ask them directly whether they are thinking about death. If they are, they need to be closely supervised and referred to professionals. Never underestimate the possibility of them taking their own life; experience teaches that clinical depression is life-threatening.

Some people argue that if someone is living in a constant nightmare, endlessly suffering and sees no point in living, then perhaps it's better not to stand in the way of their suicide. My answer to this is that it is important to understand that depression, no matter how deep, is generally a temporary phenomenon and usually passes within a year at most, even without external intervention, medicinal or otherwise. The slowdown is a type of strategy the brain employs to deal with difficulties, real or imagined. So, like a reed that bends before a strong wind but straightens when the storm has passed, ironically, the brain uses depression to help a person to survive and, after a while, it manages to return to its usual equilibrium. Another metaphor for this process might be that our body temperature increases during illness, as the body fights bacteria, then, without medical intervention, returns to normal after it has won, and the infection is under control.

As with many other processes that occur in the brain, researchers have not yet been able to give a clear explanation of how the brain manages to keep our mood in a balanced state. Although it is true that the brain can ensure people recover successfully from depression, it is not right to conclude that professional intervention is unnecessary. While it's also true that time is usually a good healer (please note that I use the word "usually" here, not "always"), in the intervening period, a cloud of suicidal thoughts is hovering over the sufferer, which puts that person at risk. It is also important to understand that proper treatment can shorten the entire process and reduce the suffering experienced.

Despite the proven effectiveness of antidepressant drugs, psychiatrists have moved toward employing psychosocial means of aiding recovery. By this, I mean psychological treatment that involves the support of relatives and friends. Brain research has added support to the idea that exposure to external stimuli and physical activity are important aids to the recovery process.

The number of people suffering from clinical depression worldwide has skyrocketed since around 1990. Neuroscientists and others in this field believe that the cause of this is the increasing pressures people feel living in an intensely materialistic society.[1] Applying what we now know from our discussion earlier in this chapter, it can be seen how clinical depression is a subterfuge that the brain uses to protect these sufferers by allowing them a temporary escape from such pressures in their ceaseless struggle for survival. The brain also works on the principle that sinking into depression will prompt those nearby to help. In modern society, this is reflected in workplaces organizing time off work and so on. It may be that, once diagnosed and, depending on their symptoms, such a break would be enough, but in severe cases, hospitalization might be required.

Postpartum depression

Also called postnatal or perinatal depression, psychiatrists recognize postpartum depression as a special type of depression, one that you may be surprised to learn is also associated with promoting the survival of mothers and has been with us since the earliest times. What we now know as a result of brain research is that the roots of this type of depression lie deep in the tribal social structure that characterized the way early people lived for thousands of years.[2] When the processing of stimuli such as nutritional status, fatigue, mental stress and physical health led a mother's brain to conclude that, in the present circumstances, she would not have the physical and mental strength to take care of both herself and her newborn, her brain would slow down its functioning and she would experience a depressive state. When the other members of the tribe recognized the mother's exhaustion, they would take it on themselves to care for her baby and, as the mother rested, so this guaranteed the survival of them both.

Despite all the social upheavals that have occurred in the thousands of years since then, this pattern of activity has survived to the present day.

Even now, when a mother's brain concludes that she lacks the strength to care for herself and her new arrival, it induces depression, which will prompt those close to her to step in. It's the same process in a different guise. The depression is the same type of depression and it's still focused on promoting survival of the mother and baby. It is simply that the environment – the social structure – is now different.

Postpartum depression can be severe, such that a mother contemplates or commits suicide, but, as with clinical depression, when the storm passes, an equilibrium will reassert itself. Since the beginning of the 21st century, there has been a marked increase in the number of mothers with this condition, and one factor that might be important is that many work. Although it might seem counterintuitive, from what we have learned so far, we can now understand that depression would be one of the brain's proven methods for helping a mother cope with fears about returning to work after her maternity leave.

Seasonal affective disorder

Another type of depression associated with promoting our survival is seasonal affective disorder (SAD). To understand this condition, let's consider the animal world and examine how the brain promotes the survival of mammals, such as bears and squirrels, during the winter months, with the intense cold and scarcity of food.

The solution the brain came up with to meet the challenges of winter is hibernation. In fact, this is a seasonal form of deep depression, and it slows the activity of the internal organs and metabolism considerably. In this state, animals can survive the difficult conditions they face during winter without the nutrition that normally would be essential. Although deeply depressed, hibernating animals are not sad, merely indifferent, emotionless. The winter sleep of bears is not continuous throughout the cold months, and so researchers who sought to take advantage of the bears' hibernation to examine them more closely escaped by the skin of their teeth. The bears, sensing a threat, shook off their depression and attacked the intruders before returning to their hibernation.

SAD is not the exclusive preserve of bears and squirrels. Evolution has conditioned the human brain in a similar way to help us survive winter too. Feelings of depression during the harsh winter months, when the possibilities of hunting and gathering were scarce, served our ancestors well, as depressed people are apathetic, think less and use their muscles less. In this state, they needed less energy, so they could make do with less food and drink.

Although modern life is completely different, it seems that this component, which is part of the brain code, has not disappeared completely. In very cold countries, such as Sweden, the rates of SAD are high. Even in warmer climates, the harder the winter, the higher the incidence of people experiencing this type of depression. As far as the light therapy suggested for those experiencing SAD goes, there is a direct relationship between the amount of light there is and the intensity of the heat of summer days. It is enough for the brain to be exposed to greater levels of light to reproduce the connection between a happier mood and lighter, brighter summer days, and so people experience a lifting of their SAD symptoms.

The phenomenon of SAD occurs in other contexts, too, such as among farmers during the fallow seasons when there is little or no work for them. In all contexts and in all circumstances, SAD is considered a mood disorder.

Mood disorders that increase brain activity

An elevated mood is a feeling of euphoria, joy, cheerfulness and enthusiasm. You might be surprised to learn that together with such emotions are also ones such as frustration, jealousy, irritation and anxiety. Remember that, as far as moods are concerned, the dividing line is not between positive and negative but between active and passive. These feelings are not so different from one another in the sense that they all cause an increase in activity in the brain. In everyday life making the transition from anxiety to curiosity is much easier than moving from anxiety to sadness.

It is among the moods found at the end of the spectrum that increase brain activity that we find bipolar disorders (formerly called manic depression), though we still do not know why this is. People with bipolar disorders experience unusual and extreme switches from one emotion to another, from high to low moods, which can happen in the blink of an eye. For example, from overwhelming joy to uncontrollable anger, and from there to a burst of jealousy or deep anxiety. This kind of emotional upheaval is out of their control. What all these emotions have in common is that they lead to turbulent brain activity involving the emotions, thought processes and behavioural impulses.

This highly active situation is the opposite of what we saw happen to the emotions during episodes of chronic depression and so on, which relate to the passive end of the spectrum and, instead, cause an inability to be active. Nonetheless, like depression, manic episodes eventually pass of their own accord.

To some extent, the suffering experienced during such episodes is not as severe as that endured in periods of depression because those with bipolar disorders are still active all the time rather than incapacitated. However, the dangers inherent in bipolar attacks are many, for both those experiencing them and others around them, because of the extreme behaviour that often results. This can include squandering money, sexual permissiveness, unrestrained gambling, alcohol and drug misuse, making threats, acts of violence and reckless driving. In psychiatry, this condition is defined as psychomotor restlessness, which means having incessant urges to be active, even if that activity has no goal, and without considering the consequences.

All this excessive activity requires a lot of energy but, during an episode, those with bipolar disorders do not feel hungry or tired, so they run themselves down to a state of complete emptiness. The inevitable result of extreme and persistent situations like this is a physical collapse. The data on manic episodes show that they are not a one-off phenomenon for people. Most will go on to have more similar episodes during their lifetime. As for the number of episodes each person has and their frequency, there is no accurate information, but it is likely that these will be combined with bouts of depression in alternating manic and depressive waves. Thankfully, in many cases, recurrence can be prevented by taking medication regularly.

Why does the brain create such trouble for people with bipolar disorders? Looking at the evolution of the brain provides some answers once again. Although it may seem unlikely, these episodes would have increased the chances of survival for our primitive ancestors and could have saved their life. During extreme events, such as territorial conflicts between tribes or confronting hungry carnivores, our ancestors had to be not only at their peak physically but also cognitively, able to ignore distractions, and all this without needing food or sleep. Manic episodes would have made all this possible and kept them alive. The implication of this is that, so long as mania serves a worthy purpose, according to prevailing social norms, it can be considered acceptable.

Since those early times, there have been people who have used this understanding of the power of mania for evil ends. During World War

II, some Wehrmacht troops, unbeknown to them, were given pep pills, which caused their brains to enter a manic state, allowing them to withstand the tough conditions on the front and fight bravely for days on end without needing sleep.[3] Also, by the end of the war, it has been found that Adolf Hitler himself was in a permanent manic state, and everything that entails, brought on by a cocktail of stimulants administered by his personal physician, Theodor Morel.[4]

Nevertheless, generally experiencing manic states is considered a serious affliction and, fortunately, we have some quite effective medicines to treat it. One of these is lithium, which is a natural mineral found in trace amounts in our bodies. Brain research has not yet discovered exactly how lithium stabilizes mood, but people with bipolar disorder who take the correct daily dose of it manage to achieve a balanced mental state. The problem is that an excessively high dose of lithium is fatal, so doctors prefer to prescribe less powerful drugs to some patients. As for depression, combining this with psychological and social support helps to ensure that treatment is successful and speeds a return to balance.

Some tips for improving your emotional state

In a situation where we, or someone close to us, experience a negative emotion from the passive end of the spectrum of emotions, such as sadness, depression, fatigue, exhaustion or loneliness, it will be easier for the brain to convert this to a positive emotion from this end of the spectrum than an emotion from the end that increases brain activity. Examples of positive passive emotions are calmness, peace, satisfaction and gratitude.

To help the brain convert the negative emotion to the chosen positive one, focus on an event from the past, during which the dominant emotion we experienced was, for example, peace. Focus on the sights, sounds, touch, smells and tastes associated with that peaceful event. Now concentrate on your inner state and relax deeply to experience peace.

Similarly, when we experience a negative emotion from the other end of the spectrum of emotions, ones that increase activity in the brain, such as anxiety, anger, frustration, jealousy, astonishment or panic, it will be easier for the brain to convert this to a positive emotion if it, too, is from this end of the spectrum, such as desire, curiosity, excitement or eagerness.

This time, to help the brain make the switch, focus on a past event in which the dominant emotion experienced was, for example, excitement.

Focus on the sights, sounds, touch, smells and tastes that you associate with that exciting event. Now concentrate on your inner state and experience the excitement.

Concentrating on such positive memories will allow the brain to copy emotions connected to that pleasant moment from the past into the present.

***

In chapter 6, we will get to know more about how wonderful our senses are. Indeed, they are the key to how our emotions function, and this relationship continues to amaze brain researchers.

CHAPTER 6
OUR WONDERFUL SENSES

If this chapter had been written in the 20th century, it would have concentrated on the places in the cortex where messages from the senses are processed. At that time, neuroscientists interested in vision, for example, would zero in on what they understood to be the part of the cortex that deals with vision. They believed that all the information coming from our eyes was sent there, identified and decoded. We could react accordingly to what it told us and our circumstances to help us survive. This was how researchers approached all the other senses as well. Thus, neuroscientists studying the sense of hearing focused on the part of the cortex that processes auditory stimuli, looking at the characteristics of signals received from the ears. Likewise, researchers looking to expand our knowledge of the sense of touch concentrated on that area of the cortex where messages from the skin are received.

This approach, which places an emphasis on the processing of information in the cortex, arose from two assumptions that were accepted at the time. The first was that the cortex is the most important part of the brain. The second was that every brain function takes place in a separate area of the cortex. Now, in the 21st century, we understand that both these assumptions are wrong.

Regarding the first assumption, leading researchers today claim that the role played by the subcortex, located deep in the brain, is more important than that of the cortex.[1] One of the reasons they give for this is that information is processed in the subcortex, which operates under the radar of our awareness and is faster than the cortex. As for the second assumption, it has become clear to researchers that, rather than functions being handled in separate areas, the brain is one big neural network. Every function that happens in the cortex directly affects other brain functions.[2] For example, a person who is feeling happy will tend to remember happy events from their past, their behaviour will be energetic and they'll feel light on their feet. A happy person's eyes will tend to register more positive images in their surroundings than negative ones. In this way, emotion directly affects memory, the focus of our attention, movement and vision. These same interactions also take place between the areas that receive other sensory information and those related to the motor and cognitive functions controlling our bodies.

Now, let's get to know our eight main senses in light of these discoveries. Yes, that's right, not five but *eight* senses.

Sight

Ask someone to name the part of the body responsible for our sense of sight and, without a second thought, they will respond, "The eyes, of course." In fact, it is the brain.

Our eyes receive rays of light and the lens focuses them so that they land precisely on the visual receptors (or photoreceptors) located in the retina, which lines the back of each eye. These visual receptors convert the pattern of light or optical message of the image on the retina into electrical signals. They do some initial processing before passing the information along the optic nerve to the brain. Once it reaches the cortex, this information undergoes more advanced processing to allow us to be able to recognize faces, objects, movements, colours and distance, all of which are necessary for the proper functioning of our sense of sight.

Until the beginning of the 21st century, brain research focused exclusively on visual processing as it occurs in the cortex, called the occipital lobe. When attention was drawn to the existence of areas in the subcortex that relate to vision, researchers and neurologists were repeatedly surprised by the phenomenon known as blindsight among people who otherwise are blind.[3] Blindsight may be experienced by people whose cortex has been damaged, as a result of trauma or a stroke, but whose subcortex functions normally. Although they are unable to see, they retain abilities that blind people who have a damaged subcortex do not have. For example, they can navigate their way around obstacles without being able to see them (see chapter 7 for more on blindness and sight).

Evolutionary approaches to understanding how the brain works have resulted in many insights. The main mission of the brain to promote survival, and all its functions developed solely to achieve this goal. Naturally, it follows that our sense of sight operates with our other senses to this end. Researchers have concluded that it is our brain which creates our reality. In other words, the brain is not interested in us seeing things as they are, only in ways that will advance our chances of survival.

To get some idea of how this works, picture someone looking for a battery. She opens a drawer and looks inside. Later, when her son asks her if she's seen his sunglasses, her answer will most likely be that she hasn't. Even though the sunglasses *were* in the drawer, she failed to notice them because an area in her cortex prevented her from being aware of the visual information about the sunglasses that she received from her eyes. As far as the brain is concerned, if we're not concentrating on something, then it's not important for our survival. In this case, as the sunglasses were deemed unimportant at the time the information about them was being processed in the cortex, the brain didn't draw attention to them.

Another important insight that brain research has uncovered is the existence of close mutual relationships between different brain functions. For example, if we gaze at a landscape familiar to us from our childhood, it is likely that we will recall childhood events related to it. Furthermore, the connections between the visual areas and other parts of the brain, including those for memory and emotion, allow us to experience the same feelings that we experienced in our childhood in relation to the landscape. It is even possible that our behaviour will also be somewhat childish.

These interactions between the different brain functions are also bidirectional. That means, while experiencing a certain emotion, the visual messages we'll be most aware of will be ones closely aligned with that particular emotion. For example, on a day when you're overwhelmed by joy, if you stand before two trees and one is in blossom, the other dying, you will probably only notice the tree that is in blossom.

Taste and smell

More than 2,000 years have passed since Greek philosopher Aristotle described the five senses known to him and his contemporaries. He also separated the senses of taste and smell. With the benefit of the findings from all the research undertaken since then, it has become clear that things are not so simple. The current understanding is that the sense of taste is the sum of the senses that allow us to characterize the chemical components of the food or other things before us. To do this, we use two senses – taste *and* smell. Therefore, for the purpose of characterizing the chemical components present in the air, we use only the sense of smell, but where food is concerned, the sense of taste is enhanced, incorporating elements of smell as well.

The tongue is able to detect a limited number of basic tastes. In the past, it was customary to recognize four tastes, whereas today we list six: sweet, salty, sour, bitter, umami and oily. However, some experts claim that humans are equipped with receptors for detecting other basic tastes, too, such as metallic flavours and that of water.

The modern world is generally one of abundance, where food is plentiful and available. This was not the case long ago, when the majority of the population would have experienced periodical food shortages. As one way to cope with times of scarcity, the brain developed the strategy of directing humans to eat as much as they could when food was available, even if they were full, just in case. It still does this. To make the strategy even more effective, the brain also directs us to prefer energy-rich foods, as these would have kept us alive for longer during times when food was scarce. The end result of this is that flavours associated with energy-rich foods arouse in us the emotion of pleasure, creating an immediate urge to eat. The flavours of such foods are sweet and fatty, and when the two are combined, we find the temptation hard to resist. Just consider the general reaction of people to the dessert menu at a restaurant. Even though we have enjoyed a hearty meal and feel sated, we will happily order a chocolate cake or some other delight, rich in sugar and butter.

Unfortunately, this default mode in the activity of our neural networks, dating back to the dawn of humankind, plays a central role in the current epidemic of obesity. Fortunately, this does not have to be our destiny. Thanks to the flexibility of our brain, we are able to reorganize the reactions usually triggered by all kinds of sugary stuff and fatty foods in a way that neutralizes the ancient inbuilt pattern in the brain so it no longer pulls us toward these foods. If you want to change your eating habits for the sake of your health, you can develop an indifference to the sight of these foods. The tools you need to do this are:

- an awareness of the harmful effects of sugars and fats on the body
- a strong internal motivation to change
- an ability to keep practising your chosen method to achieve indifference, which will likely involve gradual abstinence

Being persistent is especially useful when you get to the stage of practising, which is the hardest part of all.

One way to develop an indifference to sugary and fatty foods is to associate each food you want to avoid with an idea that causes you to reject it. Success is dependent on your motivation that is the driving force needed to make the connection possible. At first, you will need to be active in associating the food with the repulsive idea (such as vomiting or diarrhoea) but, later on, the connection will become more automatic, until you achieve your goal of an indifference to sugary and fatty foods.

Two other tastes that together cause the brain to identify foods as nourishing and important are the salty, meaty tastes of proteins and plants that are known as umami, which is Japanese for "delicious". These flavours represent two groups of nutrients that are important to the healthy functioning of the body: proteins and minerals. Among the minerals we need are sodium, iron, zinc, calcium and phosphorus.

These substances have also long been encoded in the brain as nutritious, although they are rated as less important than sugar and fat. The reason for this is that, in ancient times, having energy was the most important thing for survival, and fat and sugar are good sources of immediately available energy. Because proteins and minerals are less important to the brain, it does not link consumption of them with as much pleasure as it does for sugar and fat, but we still find them tasty in the right dose. People who are temporarily deficient in minerals such as sodium may well experience an uncontrollable urge to eat salty foods. This is because the brain recognizes the deficiency and so creates a temporary stronger connection between the taste of the missing mineral and the emotion of pleasure. This is why someone with an iron deficiency may have a seemingly irrational desire to eat soil and someone with a sodium deficiency may lick stones.

During our evolution, the brain also developed to ensure we recognize two chemical components that can indicate something is toxic or food has spoiled. These components activate the sour and bitter taste receptors on the surface of the tongue. For example, a sour taste warns us that some fruit we just tasted is not ripe, so it may contain toxins intended to deter pests. These often dissipate as the fruit ripens and, when it is ripe, it is also sweeter, so our brain then conveys that it is good to eat. A bitter taste also serves as a warning, enabling us to identify that a piece of meat is infected with bacteria. The bitterness comes from dangerous toxins that the bacteria secrete in the process of causing the meat to rot.

If I were to tell you that we've now reached the end of the list of flavours, you might disagree, thinking that I missed one. What about spicy? Whether it's chilli, curry or wasabi, you may feel sure that spicy is a taste, but let me explain why it's not. Technically, what we are experiencing instead of a taste is a reaction in response to the food activating pain receptors in the mouth.

The receptors for the six tastes are scattered on the surface of the tongue and inside the mouth, and these allow the brain to distinguish one taste from another. But our body is not satisfied with this and so it has added the smell component, creating extended taste, mentioned earlier in this chapter. This combination enables us to distinguish hundreds and thousands of foods and drinks from one other. When food is in our mouth, some of its chemical components find their way from the back of the mouth to the nose. It is this link between the mouth and the nose that allows us to inhale through our mouth and exhale through our nostrils. There, in the upper part of the nose, are our smell receptors. They absorb airborne chemicals from the food and translate the information – each characteristic of a smell having a different frequency – into unique electrical messages that are then sent to the areas in the brain that relate to smell.

For the chemicals found in food to move up into the nasal cavity, a pressure difference is required between the mouth and the nose. This is achieved by us gently exhaling air while we are eating, though it is something we are not aware that we are doing. Next time you eat, try to stop exhaling from your nose while you're chewing your food. You will be surprised to find that, when you do this, the food you had been savouring before becomes tasteless. It is not the taste that has disappeared – you will still be able to tell if the food is sweet or sour, bitter or salty – but you are missing the food's unique aroma that extends our experience of taste. This is because chemicals have not been able to make their way to the olfactory receptors in your nose, so messages you were receiving before are not being sent.

The areas of the brain relating to extended taste receive and process messages coming from the mouth to create an urge to eat or to avoid eating. Once again, all this happens to promote our survival. With this in mind, the brain also involves other parts of the cortex and subcortex. Studies have found a connection between the functions of sight and extended taste.[4] For example, the brain perceives food that is red in colour as the most nutritious and, therefore, as the most attractive. This is not a random preference. In fruit, red denotes a high level of sugar, and in meat, red signifies a high concentration of proteins. The connections between the areas of the brain relating to taste and other areas are also supported by an old saying: "Hunger is the best seasoning." We can see how, when food is in short supply, the areas of the brain concerned with taste will press the pleasure switch, causing us to devour food that we would normally ignore. We may interpret this behaviour in a rational way, telling ourselves that it doesn't taste too bad, but, in reality, it's the message that the brain has sent us that's making it more palatable.

Our sense of smell not only operates when we are taking a bite of our lunch and chewing it but also helps us to characterize chemical components floating in the air, whether they are from food or something else. Some smells, such as burning, signal danger and others signal opportunities – the aromas wafting from the kitchen signalling that we will have a delicious meal soon. Despite the immense importance of our sense of smell, which is designed to analyse these chemical particles, researchers in the past tended to underestimate its value and give more weight to the sense of sight. This preference was based on the fact that the areas in the cortex concerned with sight are much larger than the areas relating to our other senses. They account for about a third of the total area of the human cortex. Fresh studies debunk this concept. One study found that when all our senses except for smell are neutralized, our ability to follow a certain odour to its source can be comparable to a dog's.[5] Focusing awareness on smell alone provides optimal conditions for the olfactory areas in the brain to process incoming information.

Besides detecting dangers and opportunities, animals have a sense of smell to create and cement relationships with members of their own species. Smells are emitted from the animals' bodies or in the form of secretions. Females of many species have a characteristic smell when they are ready to mate. The smell is secreted by the body, carried in the air, and reaches the noses of males of the same biological species. They recognize the smell, which causes them to search for the female and try to mate with her. In many species, males secrete an odour in their urine to mark their territory.

Essentially, chemicals secreted by one animal and detected by another stimulate them to behave in certain ways. These chemicals are called pheromones. Until relatively recently, scientists believed that pheromones no longer had an influence on human behaviour, but many studies support the assertion that they do. In 1971, Martha McClintock was the first to show that the menstrual cycles of women who live communally for a long time (for example, in student dormitories, nuns' quarters or prisons) synchronize.[6] That is, they tend to all ovulate at the same time.

This synchronization has since become known as the McClintock effect. Subsequent studies have found that the trigger for this synchronization is pheromones from the women's perspiration carried in the air.[7] In prehistoric times, this effect would have been advantageous because, if all the women in a tribe ovulated at the same time, their babies would be born around the same time and they would be able to help one another with breastfeeding and child care.

The work of neuroscientist Noam Sobel has added to our knowledge about the unconscious effect smells that humans pick up from one another have on behaviour. One study found that women's tears contain a certain chemical substance.[8] When this substance was detected by the noses of men, it resulted in neural activity that modulated their behaviour. Another study found that after we shake another person's hand we tend to unconsciously smell our own hand.[9] It was concluded that we do this because our palm has absorbed the other person's perspiration, which contains unique chemicals. An analysis of these chemicals, picked up by sniffing, allows the deep regions of our brain to characterize the health and emotional state of the other person. This is all without our being consciously aware that this is happening to work out how to behave and respond appropriately, for the purpose of promoting their survival and our own.

Many neuroscientists are of the view that these and similar findings are only the tip of the iceberg. It may be that a large part of our interactions with one another is driven by behaviour derived from the brain's responses to chemicals picked up by our sense of smell.[10]

Touch

When we touch an object or another person touches our skin, the pressure exerted on the area causes changes in the skin. Pressure receptors detect this and translate the information into messages sent to the brain. The first areas of the brain that receive these messages are located in the deep brain, the subcortex. Their role is to determine whether the contact threatens our survival and to ensure we react accordingly. An acute reaction may be to get us away from whatever it is if the contact is perceived as threatening.

The touch message does not end its journey in the subcortex. It also reaches areas in the cortex, where it is processed again to identify the cause of the message accurately and pinpoint the exact location it came from on our body. Therefore, if I were to ask you to close your eyes, then I placed a pen in your hand, you could easily identify the pen by using touch only and would know that it was in your hand.

Brain research has revealed another dimension related to touch. We now know that information related to touch activates an extensive range of brain functions simultaneously, such as emotion, attention and memory. For example, studies have found that if participants hold a cup in their hand into which a hot drink is poured, feelings of tenderness, empathy and compassion toward those around them are experienced, and their behaviour will be in line with those emotions.[11] Conversely, if the drink poured into the cup is cold, the participants' feelings will also be cold. Similar findings have emerged from studies in which some participants were seated in soft armchairs and others in rough wooden chairs.[12]

It is important to understand that this is simply our default. For example, if the attitude of the participants to those around them had been cool from the outset, receiving a hot drink would probably not have caused their feelings to change to ones of emotional warmth.

During the 20th century, neuroscientists assumed that the main role that touch plays in our interactions with one another is of transmitting messages about a narrow range of emotions. In more recent years, we have come to understand that the range of emotions transmitted in this way is much wider than that. In a study that spanned several countries, actors were tasked with expressing emotions such as disappointment, anger, love and frustration, by means of touch only.[13] In the first stage of the experiment, participants were blindfolded and the actors were told to convey a certain emotion by touching each participant's hand. No words were spoken. Surprisingly, about 75 per cent of the participants in the USA identified the various emotions correctly.

The conclusion reached by the researchers was that contact enables emotions to be identified and, as a result, leads to better communication between people. In Spain, the rate in some cases was higher. This was explained as being due to the fact that in cultures where people express themselves freely in a physical way with one another, this ability is more developed than in countries where people are more restrained.

Touch is not only related to emotion, it also has a marked effect on our physical health. Research carried out in the USA has greatly contributed to expanding our understanding of this aspect of touch. One fascinating study found that premature babies whose mothers touched them while they were still in an incubator developed better and faster than babies who had not had this contact.[14]

Other research has determined that massage improves not only the muscles of newborns but also their immune system.[15] Despite the proven importance of contact for our emotional and physical functioning, in contemporary Western society uninhibited physical contact between people is not encouraged. A possible solution may lie in neuroscientists finding that touching someone's shoulder or palm of their hand is sufficient to enable optimal communication between human beings.[16] If this knowledge takes root and becomes more widely acted on, it could lead to an improvement in people's mental and physical wellbeing while keeping to accepted social conventions.

Pain

Each of our senses gives us the ability to perceive various stimuli. The stimuli can come from outside the body – via our senses of sight, taste, touch and hearing – and from inside the body. One message that reaches the brain from both external and internal sources is pain.

Pain is unpleasant, even excruciating, and most of us would rather never have to experience it. But we shouldn't be in a hurry to avoid this feeling because pain is vital to help us to survive. It is there to alert us to an injury or illness. It lets us know to pull away from something hot. It also indicates that we should move as little as possible and seek medical assistance if, say, we have fractured a bone.

Proof of the importance of pain comes from data on the life expectancy of babies born with a defect that prevents them from feeling pain.[17] On average their lifespan does not exceed six years. Their lives are short because they tend to injure themselves more frequently and more severely than is usual, with death resulting from complications such as severe infections.

When I was four I grabbed the hotplate of an iron out of curiosity. The contact with it was short because I experienced intense pain, but the event forms one of my earliest childhood memories. A toddler who is unable to feel severe pain would not be prompted to let go and would suffer more intense burns and the risk of a life-threatening infection developing afterwards.

The aim of the brain's initial processing of pain is to pinpoint the exact location of the source, its nature and the reason for it. Only in the next stage of processing does the brain create the emotional experience of pain. As the purpose of pain is to warn us of a threat to our survival, the default emotional experience accompanying it will be suffering. If the brain concludes that the situation does not pose a threat, it does not add suffering to the pain as there's no need to motivate us to act as quickly as possible. Furthermore, sometimes lesser degrees of pain may create a feeling of pleasure. For example, showering in hot water will activate the pain receptors in the skin, which will transmit a pain message to the brain. The brain's initial interpretation of this message will be to characterize the stimulus as one of a burning pain, but the more advanced processing stage – which ignites the emotional experience – will, instead, define it as a pleasurable rather than harmful experience. The reason for this is that the brain has associated hot water with having a shower. Therefore, even though there is awareness of some pain, it will not be accompanied by suffering. Instead, as survival is the expected outcome, the feeling will be one of pleasure, to motivate us to do it again.

A similar experience awaits those who decide to get fit at a gym. If they do not receive advice during their first workout, they will probably lift heavier weights than they should, set the treadmill at too fast a speed and generally overdo it. The following day, their every movement will be painful, but they will not feel like they are suffering. In this case, a pain message will reach the brain that defines its location as the muscles and skeleton. When this information is processed further into an emotional experience, the feeling will be one of pleasure because there will be the understanding that the pain is the result of physical effort rather than injury and its purpose is to promote survival so is to be encouraged.

The pain associated with childbirth may also be accompanied by the emotion of pleasure. If the pregnant woman is convinced that not only does the new baby pose no threat to her survival but it will enhance her life and that of her family, the emotional experience of giving birth may be pleasurable despite the inevitable pain of the labour. Similarly, undergoing surgery or having a procedure performed by a dentist will not cause suffering if the patient firmly agrees with the professionals that the invasive intervention will contribute to their survival.

So far, we have considered the kind of pain that results from physical damage to the body. There are two other situations in which pain is felt and accompanied by suffering but without there being a physical source of that pain. The first of these is mental pain that results from our identifying with another person's pain. Evolution has caused our brains to develop the feeling of empathy for others. If humans are empathetic, the likelihood of mutual help increases. This is a sophisticated trick of the brain because it pushes us to do everything we can to ease the pain of others. We experience an impulse to treat their pain as our own.

The other condition is chronic pain. Usually, some kind of physical damage to the body has occurred with pain and suffering being experienced. Sometimes, after a few months, healing has occurred and messages that had been coming from the damaged tissue cease, but the pain and suffering continue. This is because the neural network for pain is still busy working on the messages, oblivious to the fact that the body has healed.[18]

Chronic pain has become a popular topic for brain research, and one of the surprising discoveries has been that the pain is just the tip of the iceberg. It turns out that in those with chronic pain, a degenerative process may also occur involving all the brain's functions (motor, sensory, cognitive and physical health).[19] This is because the neural network associated with pain is connected to many other parts of the brain. The now defunct pain messages are therefore spread to other brain functions that were not involved in the original cause of pain, resulting in pain being *felt* even though there is no longer any source of the pain. The degeneration of the neural network means that the messages do not match up with the senses or where the pain was originally.

In response to the feeling of suffering that has been created, the brain slows down its other functions. If this situation persists, the brain will undo connections that it decides are irrelevant, thereby worsening the functional decline.

An effective natural treatment that can be used to relieve chronic pain is based on the understanding that one of the key factors regulating the sensation of pain is the way we experience threats or anxiety. However, before trying this type of treatment, it is vital to be certain that the pain being experienced is not the result of any real damage anywhere in the body. This needs to be determined by a qualified physician.

Feeling threatened is often what we experience when we are in pain as it is the emotion the brain gives us in response to a message received by its pain network. Suppressing that feeling can serve to interrupt the activity of the network and thereby put an end to the entire process behind chronic pain.

The first requirement for putting this theory into practice is strong motivation. People with chronic pain generally have such motivation but the brain can complicate matters. After all, the main mission of the brain is not to promote a life without pain but only our survival. Therefore, if the brain's logic is that the pain and suffering may yield some benefit that will help us to survive – such as ensuring a break from the daily grind or eliciting sympathy from people nearby – it will resist change, however motivated someone is. Therefore, this obstacle has to be overcome before the treatment itself begins. To break free, first it's necessary to eradicate the feeling of being threatened.

The intensity of a threat can be controlled by shifting the spotlight from it to some defining moments in our life. That is, to times when we were engaged in some meaningful activity and felt good about it. For example, a diver would probably recall wonderful sights experienced when exploring underwater; an amateur carpenter might remember how satisfying it felt when a cabinet turned out well; and most of us would be able to relive that feeling of achievement experienced when we learned to ride a bike or passed our driving test. The more someone can remain mentally immersed in such moments, the more the feeling of threat will recede. The goal is to use motivation and imagination to reduce its strength, which will lessen the intensity of the pain.

As the brain repeatedly reproduces such situations and re-experiences them, it will move from being in a state of threat to one of survival. This transition will, over time, cause a short circuit in the pain network. A precondition for this treatment requires confirming with a physician that there's no physical source of pain.

Balance and posture

While standing, if we raise one arm up straight in front of us while tilting our head in the same direction, we're unlikely to lose our balance. We take this for granted, but our ability to do this is the result of complex brain activity that includes sensory ability, information processing and a motor response that causes some precise contractions of certain muscles. Any permanent or temporary injury to one part of the body involved in this process will cause us to lose this ability. Moving our arm and head in the way described would result in us falling forward because the extra weight of our arm and head would not be compensated for.

To achieve a state of balance, different parts of the body work together in the following way:

- A system of receptors reports changes in the position of the body in space to the brain.
- The brain processes the messages received from the receptors.

- The brain responds with an output that instructs the muscles to contract and the skeleton to move as required to prevent a fall.

To illustrate this process, let's return to the example we started with. When we raise our arm and, at the same time, tilt our head, the brain receives a message describing the changes in the position of the body. The relevant areas of the brain process this message and respond with an output that means the necessary musculoskeletal adjustments are made to keep the body stable. Although these changes are essential to stop us falling, it is unlikely that we will be aware of them.

The receptors transmit messages about changes in our body's position to the cerebellum – an area of the brain in the subcortex that is responsible for balance and movement. Like many other animals, humans are equipped with three types of balance sensors found in three different places in the body.

- The **vestibular system** is located in the inner ear and reports on head movements.
- The **proprioceptive system** is spread along our back and limbs and reports on mechanical tensions in the joints and muscles that change with every movement. The processing of this information allows the cerebellum to know precisely, to the millimetre, where our body is in three-dimensional space at all times.
- The **visual array** is part of the retina at the back of our eyes and describes the relative position of the horizon to the cerebellum. When the horizon is level, it means we are standing or walking across a flat plane. When the horizon line tilts up, we are climbing vertically. Conversely, when the horizon tilts downward, we are descending. This information is of great importance because the balance of forces acting on our bodies is different in each of these situations, so our balance system must take this into account.

As noted, the brain is one big neural network. The connections between its various parts allow the brain to use all sorts of information that reaches the cerebellum to maintain our equilibrium. An example of this is the way the brain uses information about head movements. This comes from the vestibular system, and gives us the ability to focus our eyes on a specific object while we are moving. For example, we are able to read the lines of text in a book or on a phone or laptop screen while travelling. In this case, the brain follows the position of the head as it moves, and activates our eyes to move in a way that counteracts this movement so we can keep reading. If the head moves up slightly, the eyes will move down slightly and vice versa.

This ability enabled our hunting ancestors to focus on their prey when they needed to chase it. In a similar fashion, signals reaching the cerebellum from the proprioceptive system are used in functions unrelated to maintaining balance. These include fine motor skills and the ability to coordinate our muscles, which are necessary to perform countless everyday activities, from writing and playing musical instruments to threading a needle and putting a key in a lock.

To illustrate how this works, I will use the final example. When opening a door, the cerebellum receives continual updates from the proprioceptive system about the exact position of our hand as it holds the key. The location of the lock, of course, is stable. Therefore, all we have to do is take a good look at the lock and then the cerebellum can guide our hand toward the keyhole. After having focused our gaze on the keyhole (so the cerebellum has a fix on its exact location), it's likely that we could complete the task of opening the door with our eyes closed.

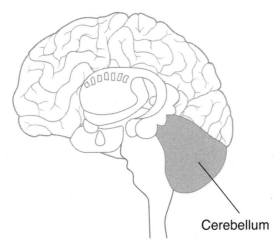

Cerebellum

After the cerebellum has used information coming from the proprioceptive sensors for the purpose of achieving balance, the information continues on its way to the cortex. There, further processing allows us to be aware of our location in space.

When we observe passers-by on the street, it is noticeable that each one of them has a different posture. One walks with a stoop, while her friend has an upright stance. This next person keeps his head lowered, while another holds her head high. Someone over there is slouching along, but her companion has his shoulders pulled back, and so it continues. Whatever an individual's posture, one thing is common to them all: they are not falling over, they are able to maintain their balance.

In the 20th century, brain researchers did not attach much importance to posture. This has changed in more recent years, with interest in the subject gaining increasing momentum. In light of the relatively new understanding that the brain is one large neural network linking completely different functions to promote survival, neuroscientists have been drawing ideas for research from holistic therapies.

One such holistic approach is the Feldenkrais method, founded by Moshe Feldenkrais, who believed that there is a close connection between our posture, general health and mental wellbeing. Indeed, this relationship has been proved scientifically.[20] For example, extroverts often have an upright posture, while introverted people tend to bend forward more with their shoulders slumped.

It is assumed that the connection discovered between the areas of the brain concerned with motor functions and those parts where our character traits are formed is there to enable humans to be social creatures.[21] In prehistoric times, the brain came to know that communal living helped human beings to survive and, to promote this, created a series of functions defined as social. One of these is the ability we have to draw conclusions about the emotional and health status of others by observing their posture.

While walking near my home, I saw a man coming toward me using short, quick steps. I also noticed he was slouched, his head was drooped and his hands limp, and I concluded he was mentally ill. Later on, I heard from one of my neighbours that this unfortunate man had recently been diagnosed with schizophrenia, and when he is not taking his medication, he often wanders the streets. I have no doubt that the ever-growing understanding we are acquiring in brain research about the close connection between posture and mental functioning will come to serve the field of psychiatry for the purposes of diagnosis and in the field of care.

As noted in chapters 3, 9 and 11, knowing that we have a flexible brain presents medical professionals with new options for moderating how the brain functions without recourse to prescribing medicines. This is also the case when it comes to posture, and our ability to use it to discern and influence character traits and mood. The relationship between posture and character is bidirectional – that is, posture is affected by character traits, and character traits are affected by posture. This understanding has practical applications because by correcting posture, improvements in character traits are possible. For example, if a person who stoops learns to adopt a more upright posture, it is likely that they will also enjoy greater confidence in themselves. Improving self-confidence will reduce their feeling of being threatened and may lead to them becoming more proactive. Anxiety and feeling threatened are known to raise levels of cortisol in the blood, which is a response to stress that is harmful, so by reducing these negative emotions, which will also lower cortisol, both physical and mental health can be improved.

Hearing

Our sense of hearing assists and affects us in three different ways:

- It enables us to recognize voices and sounds.
- It gives us information about the location of objects around us, contributing to our spatial awareness.
- It affects our emotions and feeds into impulses for automatic behaviour.

The wonder of being able to hear is made possible by our ability to convert air moving around us into sounds. These movements create sound waves that enter the ear and pass along a channel known as the auditory canal to the middle ear, where they cause the eardrum (or tympanic membrane) to vibrate. The vibrations move three auditory bones, known as the hammer, anvil and stirrup (or malleus, incus and stapes), which amplify the sound. In the next step, the waves reach the complicated structures of the inner ear (the vestibule, semicircular canals and cochlea), which translate the sound waves into electrical messages. Next, the messages travel along nerves to the brain, where they are decoded and translated. We can then hear the sounds.

Sound waves differ in terms of their speed (the frequency of the wave), which is measured in hertz (Hz), and intensity, measured in decibels (dB). Our hearing system is somewhat limited in that we are only able to process frequencies in a range between 20 and 20,000Hz. If anything is outside this range, we cannot hear it at all. That is why, although we are continually surrounded by sound waves, we are not conscious of the vast majority of them. This includes infrasound waves (with a frequency lower than 20Hz) and ultrasound waves (above 20,000Hz).

Our ability to perceive sound is also limited in terms of volume to a range between 10 and 150dB. This is because it is difficult to translate sound waves accurately into electrical signals when they are lower than 10dB and higher than 150dB. Also, extremely loud sounds can cause damage and result in hearing loss.

The brain initially processes electrical signals from the ears in areas at the sides of the cortex. More in-depth processing happens in adjacent parts of the brain. This is where sounds might be recognized as language or for their unique meaning. We remember what they are, such as a cat howling, a truck reversing or a piece of music.

Some people are gifted with an exceptional ability to process the sounds of voices and tones. This ability is classified as either an enhanced capacity for relative hearing or absolute hearing. Enhanced relative hearing allows someone to recognize and remember the different frequencies between sounds. Almost all of us are able to recognize that there is a difference between one sound and another, but a person with this ability who plays a musical instrument will be able to play any melody they know without a score because, over time, they will accurately recognize the frequency intervals between the notes.

Unlike relative hearing, those with enhanced absolute hearing can identify and remember every individual sound. Most of us will remember a certain sound or a new melody for a short period of time. A person blessed with a high level of absolute hearing, who knows the names of the notes in the various musical scales, will know the note for each sound.

Further processing of the auditory information enables us to identify where voices or sounds are coming from. When someone's phone rings during a lecture, I am aware of what it is that makes the sound I hear and the sound's meaning, but my brain is capable of giving me much more information. I can locate approximately where in the room the sound came from and sometimes whose phone rang.

This is possible because we are blessed with two ears. Due to the distance between them, the sound of the phone ringing is received with a different intensity in each ear and at a different time. The auditory messages sent to the brain by each ear are therefore different. When they reach the areas in the brain where they undergo integrated processing, these differences enable us to locate the sources of the sounds in space. If the sound of the phone ringing is louder in my right ear, my brain will understand that the source of the sound is on my right.

Our earlobes also have an important role to play in helping us to locate sounds. Voices and sounds coming from different directions hit different parts of each earlobe, creating subtle distinctions between the components of frequency and intensity of the sound waves entering each ear. These slight differences between the auditory stimuli help us to work out where sounds are coming from quite accurately.

Our sense of hearing serves us faithfully in our day-to-day life, and was even more essential to our prehistoric ancestors. Ranking much higher for them than being able to recognize voices and where sounds were coming from was the information hearing provided about what produced the sound they could hear, and the guidance on how to react that the brain gave in response. The frequency and intensity associated with the sound of a sabre-toothed lion's roar would have activated the emotion of threat and stimulated an automatic response to behave in a way that would give our ancestors the best chances of escaping. Once again, the goal was survival. In contrast, the bellow of an antelope would have automatically triggered a pleasant emotion and the response of chasing after it in the hope of a hearty meal. Our abilities have adapted over time to serve us effectively in different ways in the modern age. The sabre-toothed lion's roar has been replaced for some unlucky people by their boss's voice, while the antelope's bellow has been superseded for many more of us by the tune played by an ice cream van.

Synaesthesia: confusion of the senses

After a lecture I gave about vision, a young woman approached me and said, "You'll be surprised to hear that when I think of a specific month, I see a certain colour before my eyes. Each month has a different colour. June, for example, is always red."

I wasn't surprised at all. Synaesthesia is a familiar phenomenon. What happens is that a real stimulus coming from one sense, such as hearing a voice, activates a response in the brain from an area that relates to one of the other senses, such as seeing a colour. After all, the brain is one gigantic neural network, so any stimulus has the potential to be received and processed in any part of the network. This doesn't usually happen as it would hinder communication between people. For those who do experience synaesthesia, there are fewer of the natural restraints that limit the movement of messages relating to inputs through the neural network.

There are something like 50 varieties of this type of confusion. The most commonly experienced combinations are between language and colour, music and colour, and language and space. The phenomenon affects about 4 per cent of the population around the world and is not considered to be the result of a disease or disorder. For about 50 per cent of these people, the confusion they experience is between language and colour. This ratio is the same for both sexes.

People with synaesthesia think that what they are experiencing is the norm as it feels perfectly natural to them. Some neuroscientists maintain that before language developed all humans were endowed with this confusion of the senses, as it was another effective way to ensure that our ancestors survived.[22] The argument goes that if hunters regularly saw a certain colour in front of their eyes or experienced a sour taste in their mouth when they heard the growl of a wild animal, for example, this would make the warning of danger clearer to them than a response from one of the other senses.

The cognitive or linguistic revolution of the development of language, which took place about 70,000 years ago, turned this advantage into an obstacle. When humans could express in words the sensory stimuli and outputs of their brains, it was important that things were perceived in a uniform way by everyone in the tribe. If one member of the tribe claimed that the growl of a wild animal was experienced as a red colour, while another member recognized that same growl as having a sour taste, there would be a breakdown in communication, causing confusion. This would harm their chances of survival, and the future of the entire tribe would be weakened. According to this theory, the brain developed in such a way as to inhibit what had been a natural ability to mix the messages of the senses.

So for a small section of the population, sensory confusion has survived until the present day. Within all of us, there are some residues of this phenomenon. One study has clearly demonstrated this by presenting participants with images of two shapes.[23]

The participants were told that the nicknames for the shapes were Bubba and Kiki and asked to match each shape to the nickname they thought most appropriate for it. At least 95 per cent of the respondents, who were from a variety of different cultures, gave the shape with the soft outline the name Bubba, and chose Kiki for the angular shape. The researchers concluded that this stemmed from a process in the brain in which hearing and vision are combined. This made the sounds in the name Kiki appropriate for a structure with angles and Bubba for a round shape.

To understand what the experience of synaesthesia is like, here is some correspondence from a girl who responded to an article on this subject.[24]

I was born and raised with synaesthesia and never knew it was anything unusual until a few years ago. As far as I know, there is no other person with synaesthesia in the family (I even asked if there had been someone in previous generations who passed away). When you synthesise, you just accept it. It's not something you can fight, it just happens naturally and with clarity.

My synaesthesia expresses itself in colour–letter linkage. Therefore, every letter I see is expressed as a colour in the brain and vice versa, every colour I see I convert into a letter. I don't make spelling mistakes because each word has a very clear sequence of colours in my mind. Sometimes, the clothes a person is wearing create words for me. I can remember entire texts by heart and one reading is usually enough for me to absorb all the information I need, especially dates and names.

My synaesthesia, as I see it, is a great advantage in various areas of life. My memory is better and I make zero spelling mistakes, I can quickly learn a new language and I have a good grasp of mathematics. I finish word puzzles in a minute and find words other people never see. It's a shame that synaesthesia is not familiar to most people. I believe we might find more synesthetes who are not aware of their condition. I didn't know I was different until recently when by chance I was exposed to the phenomenon. Please carry on spreading the word and explaining all about it.

A tip for improving your sense of touch

Touch is powerful, so a hug has great potential to bring people closer together and foster bonds between friends, spouses, parents and their children, and so on. It also serves as a mediator of emotion between people. According to the intensity of the hug as well as its duration, our subcortex will draw conclusions about the emotional state of the person hugging us. The same goes for the person receiving our hug.

Identifying the other's emotional state allows for mutual expressions of empathy between friends and spouses. When parents and their children hug, the situation is completely different. The parent who, filled with sadness, worry or anxiety, hugs their child simply to fulfil their parental duty might be doing more harm than good. The parent's negative emotions will be received by the child's senses, conveyed to their neural network and could have adverse physical and mental effects as the parent will not have conveyed the emotion of love to their child.

I am not suggesting that any child should be denied hugs, but it is worth ensuring that when we hug them, we are experiencing a positive emotion because that is what they will receive. Consequently, it's a good idea to take a few seconds to concentrate on some significant and happy event before enveloping a child in a hug as, then, the pleasant emotion we feel will be passed on to the child.

***

My father, who was exceptionally wise about life, told me many times that, with all due respect to the senses, they may deceive us. In chapter 7, we will understand how right he was.

CHAPTER 7
HOW THE BRAIN CREATES REALITY

"I think, therefore I am."

The ideas in this famous statement, written by 17th-century philosopher and mathematician René Descartes in his *Discourse on Method* in 1637, have inspired generations of neuroscientists. Understanding that the senses may deceive us, Descartes concluded that self-awareness and thinking are the foundation of human existence. In this chapter, we will get to know these two components that contribute so much to the functioning of the brain and find out how they dictate the way we go about our daily lives.

Consciousness: what is reality?

Reality refers to the state of things that exist as they really are. Neuroscientists have found that the brain creates our reality and the frontal area of the cortex – the prefrontal cortex – processes something as a reality rather than something we've imagined, a thought or a memory. This kind of brain activity happens following the instructions given to it by the brain code written millions of years ago. It is not impossible for the labelling to be wrong, which may manifest as false memories, delusions and hallucinations.

Awareness of our environment

When we talk of consciousness, we mean that a stimulus from the outside world has been processed in the cortex and the brain has allowed it to enter our awareness. This process, despite its great importance, is not a necessary condition for life. The decisive proof of this is that fish and reptiles lack a cortex but have been surviving well without consciousness since they first came into existence. So, how does a little fish manage to escape from a fearsome shark without being aware of the danger lurking nearby? Let me explain.

When driving out of a car park, a camera at the exit recognizes your car's number plate and transfers the details to a computer. If we have paid the parking fee, the barrier will automatically rise in a process that is a combination of inputs and an output. The inputs are the number plate and confirmation that we have paid, while the output is the lifting of the barrier. The computer entrusted with this task is not endowed with self-awareness, and nor is the little fish. The fish's eyes receive the information (the dangerous shark approaching), transfer it to the subcortex, where it is processed, and the output (an attempt to avoid the shark) follows. All this takes place without any consciousness being involved.

You may recall from chapter 1 that in the course of evolution a new area – the cortex – developed in many animals, including humans. As the brain continued to evolve, consciousness was created. Initially this was of no interest to the brain, whose goal is to ensure we survive and reproduce. But it became clear to the brain that consciousness increases our chances of survival. The result is that the areas of the cortex that enable consciousness have not replaced those in the subcortex where it is lacking but have developed *in addition* to them.

This has created an interesting situation, one that characterizes all sentient beings whereby the brain is continually operating on two parallel tracks. Every stimulus from the environment received by one of the sensory organs is processed in two brain areas – the subcortex and the cortex. Although we are aware of the information processed in the cortex, it is important to understand that, at the same time, there is a rich accumulation of information arriving from the sensory organs and being processed in the subcortex that we are not aware of at all.

We know about this duality because of evidence from clinical records, such as what happens when the area in the cortex that processes visual messages has been completely destroyed, but the eyes, optic nerves and the part of the subcortex that handles visual information remain intact.

Although the experience of people in such a situation is that they cannot see and often need assistance via a cane or guide dog, functionally, in certain ways, they can. You may remember from chapter 6 that the medical name for this is blindsight.

In a video that shows the kinds of abilities those who are blindsighted have, a man stands at one end of a corridor strewn with various kinds of obstacles, including a chair, a stool and a lamp.[1] The man is asked to walk toward a doctor at the other end of the corridor. The only instruction the man is given is to bow his head as though he were looking at the floor.

Amazingly, the man walks around the obstacles and completes the task. When asked how he managed this, the man replies that he simply doesn't know, but says, "I felt urges to move to the side, tilt the body, lift a foot . . ." He was not aware of the obstacles, so how did he manage to walk around them?

The answer lies in the area in the man's subcortex that deals with visual messages. It received messages from stimuli registered by his undamaged eyes, processed these and then activated the correct muscles to prevent him from hitting the obstacles, as doing so could have endangered his survival. All this was done without his being aware of it, in the same way that an unconscious impulse enabled the fish to escape the shark.

To overcome situations such as how to walk down a corridor full of obstacles, the subcortex retrieved information stored in its deep memory networks. An excellent illustration of how this helps both the blindsighted and those who can see is this amazing story of an American fire chief, which attracted a lot of interest from brain researchers. The chief was attending a house fire when he suddenly directed his crew to drop everything and evacuate the house. The firefighters obeyed his instruction and a few minutes later the wooden floor on which they had been standing collapsed and sheets of flames shot up from the basement.

When asked by journalists how he could have known what was about to occur, the chief stammered, "I have no idea, I just had a feeling that something was going to happen." But the findings of brain research can give us a clearer answer, which is that the chief had experienced similar situations before. The particular combination of the temperature, smell, smoke and sounds that his senses had registered before had been stored in his deep memory. When his senses detected a similar combination and the messages reached the deep areas of his brain, it triggered an automatic urge in him to flee the scene. What appeared to be a reaction based on gut feeling rather than knowledge was, in fact, the result of a careful analysis carried out by his brain without him being aware of it.

These examples raise the question: out of the countless number of stimuli that our senses pick up from our environment, which ones will be processed in the cortex rather than the subcortex and make it into our consciousness? The answer is quite simple: only information the brain deems important for us to be aware of reaches our cortex and only then do we become conscious of it.

According to neuroscientist David Eagleman, only a fraction of the enormous river of information washing over us reaches our consciousness, despite the fact that most of it does arrive in the subcortex.[2] This is similar to our knowledge of current events. Myriad events are happening all the time, all over the country we live in, all over the world, but we are only aware of the ones the media tells us about after sifting through and editing information from the massive amounts available.

In the sorting process carried out by the brain, emotions play a central role, as we saw in chapter 3. Events that are picked up by one of our senses and then trigger an emotion are considered relevant to our survival, so they are brought into our awareness. Similarly, when we decide to focus our awareness on an idea or an object or when the brain decides to do so for the purpose of promoting survival, the neural network in the cortex that is in charge of concentration is activated. A curtain comes down to keep everything else that is coming in hidden. We will only be aware of the specific matter we are focused on, and other messages reaching the brain about stimuli are ignored.

An experiment conducted by psychologists Christopher Chabris and Daniel Simons proved this very clearly.[3] Participants were shown a short film, in which two groups of people played with a ball. They were asked to count the number of times the members of one of the groups in the film passed the ball between them. At the end of the film, they were asked to share their results. Most of the scores were accurate.

"Well done!" the researchers said, praising them, adding, "But maybe you noticed something else?"

More than half the participants in the experiment – which was carried out with multiple groups – answered unequivocally, "No." Remarkably, they had not noticed that during the game a man dressed as a gorilla walked in from the right side, stood in the centre, surrounded by the players, beat his hairy hands on his chest, then walked off the left side, out of shot. In later versions of the same experiment, significant other details were changed in the film shown, such as the colour of the background and one of the players disappearing.[4] Most viewers did not notice these things either.

Illusions about reality being created by the brain is not a new discovery. As we discussed earlier, René Descartes recognized this back in the 17th century. More recently, neuroscientist Chris Frith expressed it well in his catchy phrase, "The mind creates our reality."[5] The brain presents us with a reality that it believes will best support our survival. When we're focused on something specific, such as the number of ball passes, the brain concludes this is the most important thing for our survival at the moment. It ignores all other stimuli, such as the gorilla, that our senses register in the same time period.

For the sake of promoting the supreme goal of survival, the brain even decides *how* we should see the everyday world around us. This is illustrated in a video of a Charlie Chaplin mask on a rotating pole.[6] When the front of the mask is before us, we see Chaplin's face. When the back of the mask is revealed to us, we should see the hollow inside of it, but we don't. The inside of the mask looks just the same as the outside – that is, coming out toward us rather than receding, albeit without the colours.

How is this illusion created? A frontal area in the cortex operates on the understanding it has, from the algorithm in the brain code, that a person's face cannot normally appear hollow. With the intention of making it easier for us to prioritize our survival, the brain neutralizes the message received from the eyes, converting it into a virtual reality message. So we see what we would normally expect to see if we were looking at a face. In the event of a conflict, areas of the cortex responsible for the higher-order functions in the brain will dictate what we *should* be seeing, and we don't see what is really there.

What is at work in these two examples plays out in all sorts of ways in our everyday lives. For example, because we do not concentrate on routine actions this tells the brain that they are not important for our survival, so information from the senses about them will not be brought into our awareness at all. This is why we fail to remember whether we locked the door or took our medicine.

I will tell you a story now about a slightly embarrassing experience I had that is an example of the brain creating our reality as a result of a routine. There was a canteen where I worked and I would often go to it for a meal, choosing a salad displayed at the end of the counter. One day, I was in a different canteen. I looked for the salads in the usual location but couldn't see them. One of the people working there, who probably noticed me looking at a bit of a loss, asked me if I needed any help.

"Thanks," I said. "Do you not have any salads today?" The man looked rather surprised, then said, smiling, "Turn round!"

I did and, sure enough, a few steps away was a table loaded with salad bowls! It was now clear to me that, on my way to the main food counter, I had walked straight past all those salads. They should have been impossible to miss but, as I hadn't expected to see them there, my mind hadn't bothered to move the visual information into my consciousness.

Studies that have looked into the effects of hallucinogenic drugs shed light on the mechanisms in the brain that are responsible for creating awareness. People report strange, otherworldly visions after taking such drugs. In the past, it was commonly assumed that hallucinogenic drugs distort normal brain activity but, more recently, the assumption is becoming accepted that they neutralize a visual censor in the brain. Paradoxically, then, it would appear that what people high on hallucinogens see may be closer to reality than what people who are not high see. Therefore, if a person under the influence of hallucinogens were to watch the rotating Charlie Chaplin mask, they might see it as it really is.

Self-awareness

So far, we have looked at one level of consciousness reflected in the brain's ability to bring to our attention the existence of messages concerning stimuli from our environment. A higher level of consciousness than that is self-awareness. The ability each of us has to distinguish ourselves from the environment we are in and from all the other individuals around us. In other words, our ability to be aware of the fact that we are unique and different, separate from the world and other people.

Each of us is made up of billions of the basic units of life: cells. However, in our mind, we are each one indivisible unit. Self-awareness allows us to acknowledge that we have feelings, thoughts and memories. This type of self-awareness – introspection – gives us humans an advantage in everything related to survival. We can understand our needs, find ways to achieve them and apply ourselves to a task. The brain has found a wonderful way to coordinate the different types of awareness: consciousness, which is awareness of the outside world, and introspection, which is awareness of the self.

Rafi Malach and his colleagues at the Weizmann Institute discovered an example of such coordination. They found that when our brain is aware of external stimuli – sights, sounds, smells – this awareness inhibits our ability to engage in introspection.[7] For example, during a morning run, a person can be in a state of self-observation – examining how fatigued they are, calculating how much time is left before the end of their chosen route or thinking about the vacation they are planning or even the meaning of life. But if they then started listening to music, the brain would inhibit the self-observation that had been going on. Our energetic runner would, instead, be aware of the music, the various thoughts about fatigue and so on disappearing from their mind.

Another example is that of people talking loudly on their phone in a public place. Their awareness is focused on the voice they are hearing, which paralyses their introspection. Without the censorship of introspection, they do not realize that they are speaking loudly, sometimes even sharing personal information they would otherwise not talk about in front of complete strangers. Similarly, talking on a phone while crossing the street can block awareness of imminent danger, as traffic accident statistics will testify.

As with any of the brain's functions, there is a risk that the mechanism that brings things into our awareness will not work properly at times. One very well-known malfunction is that of hallucinating. During a hallucination, a person may feel certain that they can see someone walking toward them, but the street is completely deserted. They may also hear a voice talking to them, even though no one is around. Such experiences may indicate that there has been a collapse of the higher brain functions associated with processing external information.[8]

Another known fault connected to self-awareness is that people who have experienced mental trauma may find themselves unable to remember certain episodes from their life story. This does not mean that those memories have been erased, but there is a delay put on their retrieval. I participated in a study conducted in the brain research department at the Weizmann Institute and we located an area in the cortex that, when activated, prevents the retrieval of memories.[9] As we are not talking here about erased memories, it is possible to overcome this type of malfunction, but this needs to be done with the help of targeted psychological treatments.

What is sometimes called a split personality, or dissociative identity disorder, is the result of a malfunction. It is an exaggeration of the natural brain activity involved in self-awareness.

Several decades ago, Walter Mischel, a neuroscientist, argued that our personality is not made up of a single layer, but is an aggregate of several subpersonalities.[10] Any one of these can become dominant under different conditions, enabling us to modify our behaviour to suit a situation. For example, how we behave when we are with family may be quite different from when we are with friends or at work. In one situation, we will be serious, in another, light-hearted. We may be extravagant in one setting while, in another, we are balanced and work things out carefully. As you can imagine, the whole range of possibilities is immense.

In a healthy person, what we might call a versatile personality is well-regulated and adapts to the circumstances they are in. In contrast, someone with dissociative identity disorder distorts this balance so that what should be facets of a single personality divide into one or more other personalities, each with a separate and independent consciousness. If we imagine our personality as a soap bubble, the inner walls divide to become several separate bubbles.

A healthy person caught in a threatening reality will muster the mental strength needed to deal with the situation by bringing out the appropriate elements in their personality. However, someone with dissociative identity disorder facing a similar threat might call up a subpersonality that does not fit the situation at all.

To demonstrate how this works, imagine that a healthy person goes to study traditional Chinese medicine in Beijing. Loneliness, language difficulties and an unfamiliar culture are all difficult for her, but she mobilizes the best positive aspects of her personality to cope with the situation. She has the ability not to take things to heart, thinks rationally and is focused on the positive results of her studies, so she copes successfully with this experience.

A person with dissociative identity disorder would handle this situation completely different. One outcome might be that the threatened brain would opt for outward aggressiveness, on the basis that this is the best way for it to survive. For the same reasons, the brain might also call on an obsequious personality, with the intention of getting closer to those around them and gaining their affection. This would find expression in the creation of a separate system of awareness with uniform and rigid personality traits. From this, we can conclude that malfunctions related to consciousness are the brain's way of dealing abnormally with normal, everyday situations.

Thinking

Now that we are familiar with the various aspects of consciousness, it is easier to understand the other important brain function that Descartes referred to – thinking.

If I ask people to tell me what thinking is, most of them will mention learning, analysing problems, calling up memories, planning moves, drawing conclusions and similar things. This is all true, but they are not the whole picture. Thinking – in its broadest sense – is the way the brain organizes and processes the messages picked up by our senses. Messages are received simultaneously in parts of the cortex and subcortex, but the ways each of these areas process them may be different. Despite this duality of data processing, and the fact that the results might be different, the goal is the same: to ensure our survival.

Consider the following. Your boss is critical of some work you have done. The stimuli relating to this event go through parallel processing or parallel thinking. The subcortex might process the information as a threat, so respond with defiant aggression. In the cortex the result for the very same stimuli might be to show some restraint and even to try placating your unhappy boss. These reactions are pulling in different directions, so what would you do?

One of the insights of eminent scientists such as Daniel Kahneman, and the examples of David Eagleman and Dan Arieli, is that thinking in the subcortex is immeasurably faster than it is in the cortex. Kahneman described it well in his bestselling book *Thinking, Fast and Slow*.[11] In the scenario of the critical boss, we might assume that the quick reaction of the subcortex is also the dominant one, but this is not the case. As luck would have it, in the prefrontal cortex there is a part called the conflict area. As the name suggests, its role is to identify conflicts between the results of processing in the subcortex and in the cortex and serve as an arbiter between them. When the conflict area is working properly, the clash of interests that may occur behind the scenes will not reach our consciousness and the brain will adapt our response to suit the situation. In the case of the critical boss, it is likely that a normative person (one whose behaviour is generally aligned with the norms of society) will react with restraint, totally unaware of the aggressive option the subcortex was probably suggesting. However, even in the case of normative human beings, it sometimes happens that the arbiter is having an off day. When we are hungry or tired, say, the arbiter may miss its cue to step, and so the subcortex will determine how we react.

It is customary to divide the processes that occur in the cortex into lower-order and higher-order thinking. We share the attribute of lower-order thinking with a significant number of animals that have a cortex, but higher-order thinking is unique to us humans, and it is thanks to this that our species is the most dominant of all the species on Earth.

However, we should not underestimate the capabilities of lower-order thinking as it makes important contributions to our survival and that of other animals with a cortex.

Lower-order thinking

The components of lower-order thinking are the organization, sorting and processing of information that reaches the cortex. An example of the process involved is the steps that lead to a dog being able to locate a hidden object that it has sniffed before.

1 The dog receives some initial information – the smell of an object.
2 The smell is stored in the dog's memory.
3 Faced with a multitude of smells in the environment, the dog can identify the smell of an object as being the same as that in its memory

We humans use similar processes to help us in our daily lives too. For example, they come in handy when we're scanning a crowd to find a friend we're meeting. The information we have in our memory about their appearance – face, hair, build and so on – allows us to pick them out from the other people there.

Higher-order thinking

As mentioned, higher-order thinking is unique to humans, and it takes us beyond the limits of the here and now. When the brain applies the skills involved in this type of thinking, the raw stimuli received from our senses go through two types of processing. The first is the processing of the messages from the senses, which is lower-order thinking. The second, which is the distinctive feature of higher-order thinking, is the processing of the messages from the senses together with our relevant prior knowledge of other materials that exist in the brain's neural network for memory. In this way, we can draw on our past experiences, even if those materials come from completely different fields, and move backward and forward in time. Thus, we can consider things not only in the present but from the past as well. We call this kind of thinking creative thinking.

Drawing conclusions, critical thinking and creative thinking are three types of higher-order thinking that we are all familiar with, but how do they work?

Drawing conclusions

The way we operate in our daily lives is largely based on the process of drawing conclusions, so we need to delve further into this type of thinking as it is helpful to understand it better.

"The Fox and the Vineyard", a Jewish fable from the sixth century, is based entirely on drawing conclusions. A hungry fox passed by a vineyard full of ripe grapes and wanted to get in, but it was fenced off. The fox walked around the vineyard until he found a hole in the fence, but the hole was too small for him to squeeze through. What did he do? He fasted for three days until he was thin enough to pass through the hole. The fox then spent three days in the vineyard gorging on grapes until he grew fat. So fat that he couldn't get out through the hole. His solution? He repeated the three-day fast until he was thin again, but left the vineyard hungrier and thinner than he'd been in the first place. The fox turned his face to the vineyard and lamented, "Vineyard, vineyard! How fine you are and how beautiful your fruits are, but I could not enjoy them. I came hungry and left hungry."

What happened here? The sensory message received by the fox (that he was too big to fit through the hole in the fence) was combined with the relevant knowledge stored in his memory (fasting for three days would allow him to lose weight so he could navigate the hole). The conclusion he drew was that he had to repeat the method he used to get into the vineyard to escape from it, thereby not achieving his goal.

This is indeed a witty parable with a sophisticated, although clearly unrealistic, moral at the end. Unrealistic because foxes, like most other animals, do not have the ability to draw conclusions. That this is true is clear from the premise on which the often-cited monkey trap is based. Set by some hunters long ago in Africa and South India, this involved drilling a hole of a certain size into a coconut and, after emptying its contents, putting in nuts, grain or rice, which monkeys love. Then the hunters would retreat and lie in wait for one of the monkeys to take the bait. Sure enough, a monkey would usually come and stick its hand through the hole in the coconut. After grabbing a handful of the treasure inside, it would try to pull its hand out, but wouldn't be able to and it would think its hand was stuck. Because monkeys are not capable of higher-order thinking and drawing conclusions, the monkey in this situation wouldn't be able to use the memory of how it got its hand in there in the first place and then think to simply reverse the actions and let go so it could escape.

Humans are unique due to the linguistic revolution of the development of language, which took place about 70,000 years ago and accelerated our ability to think. Thanks to language, we are able to express abstract ideas beyond the here and now. Language gives us the ability to recall past events from memory and apply them to current situations.

Critical thinking

Another type of higher-order thinking is critical thinking and this can be defined as an objective assessment of data whenever possible. To evaluate current data, we must compare the information with relevant knowledge gained in the past. For example, after a reviewer reads this book with the intention of writing a (hopefully positive) review of it, they will naturally use their higher-order thinking abilities. To do this, it is not enough that the reviewer reads and understands the book. To come to a proper assessment of the quality of my writing, they must also compare the information contained in it and the way it is presented with similar books that they have read before.

Creative thinking

Western culture in the 21st century places a lot of emphasis on using critical thought and thinking to draw conclusions. However, the most significant scientific, technological and educational achievements are the result of creative thinking.

These different types of thinking are characterized by the different ways in which the brain processes the stimuli that reach it. In creative thinking, the way data are processed is multidimensional and multidirectional. In brain research, it is customary to define this as lateral thinking. The brain uses this process in an effort to overcome a difficult obstacle: our mental blockages. The reason we have these is that the brain, even in Western culture, does not want us to be creative. It is only interested in us doing things that will mean we survive with maximum efficiency. The brain has therefore developed data-processing templates (stereotypes), so we can do this faster and with less effort. But we do pay a price for this. At best, it manifests as a lack of creativity and, at worst, as mistakes.

The findings of an extensive study provide some instructive proof of the fixation we have with stereotypes. The researchers put the following question to a group of American adults.[12]

An individual has been described by a neighbor as follows: "Steve is very shy and withdrawn, invariably helpful but with very little interest in people or in the world of reality. A meek and tidy soul, he has a need for order and structure, and a passion for detail." Is Steve more likely to be a librarian or a farmer?

Statistically speaking, the chances are that Steve is a farmer, simply because there are way more farmers in the USA than librarians. In addition, the way he's described also fits with how a lot of farmers are. Nevertheless, around 90 per cent of the participants decided that Steve is a librarian because thinking in terms of stereotypes is an obstacle standing in the way of deep, higher-order thought, and it is that which was followed here.

Each of us has a natural ability to process data creatively but, to do this, we must ask the brain to switch temporarily to a form of data processing that it deems inefficient. We need to shift our brain into a state in which it will combine the messages concerning stimuli currently coming from the senses with the ever-increasing numbers of scraps of information that it has accumulated in the past. Studies have revealed that there are several simple ways to bring about this change in brain activity. These include daydreaming, meditation, brainstorming and even, to a certain extent, limiting the resources we have at our disposal.[13] A good example of the effectiveness of limiting resources can be seen in the results of an interesting experiment conducted by Tina Seelig at Stanford University.[14]

At the beginning of the semester, Seelig gave one group of students $100 and asked them to do their best to make a profit over the weekend. She also made the same request of another group of students, but allocated them only $5. The first group bought some cleaning materials and washed cars, making a profit. The students in the second group invested their $5 in phone calls to local restaurants (this experiment was conducted at a time when each call was charged for), reserving tables for their peak times. The students later arrived at the restaurants and sold the reserved tables to people who hadn't phoned ahead and were queuing to eat. The students turned their $5 stake into a profit that far exceeded the money made by the car washers. Seelig repeated this experiment several times, with similar outcomes.

The explanation for this is that to complete a task with limited resources, the brain needs to switch from its usual patterns and stereotypical thinking to creative thinking. Another example of the liberating power of limitation comes from my illustrator friend, Dani Kerman.

While teaching at the Bezalel Academy of Art, Kerman and I conducted some workshops for the general public to develop their creativity, and Kerman tasked the students with drawing a cartoon on any topic that came to their minds.[15] After most of the students had difficulty with this, Kerman asked them to write the name of a character and then, next to it, two nouns of their choosing.

"Now," he said, "draw a cartoon related to these three elements." Most of the students, Kerman told me with a broad smile, completed the second assignment much more easily than the first. Not only that, they were also very creative.

Some tips for improving your creativity

The processing of information in the left hemisphere – the part of the brain where, for the most part, stimuli relating to language are received and processed – follows a rigid, structured format, which is the opposite of creative. An important way to promote the creative processing of information would be to either send verbal information from the left hemisphere to the right hemisphere or to the subcortical areas of the brain, located deep inside it.

There are several strategies we can use to make the brain allow this transition to happen. Yet, before implementing the strategies, the brain will need to understand that the problem we are wanting to find a creative solution for has significance. According to the brain code, concentrating on the problem for several minutes (not on the solution) will establish the understanding that this is a significant issue in the brain's networks. Think of a problem and focus on it.

Now we can become familiar with the seven strategies that will allow the verbal representation of the problem to be transmitted to the non-verbal areas of the brain. This area is where the creative processing will take place.

1 **Daydreaming** – Focus your eyes on a certain object (a picture, a flower, a glass of water) and soon you will find yourself staring but disconnected from the object. Continue for several minutes.
2 **Meditation or hypnosis** – Reach a state of gazing by following a guided recording for relaxation.
3 **Eureka** – Before going to sleep, ask yourself – concentrating while you do this – to dream of a creative solution to the problem.
4 **Brainstorming** – Together with several collaborators, jointly come up with thoughts about and possible creative solutions to the problem.
5 **Limit your resources** – Take as inspiration that you prepare your most

creative meals when there are hardly any ingredients in the fridge.

6 **Dress differently** – According to the brain code, if we dress like Indiana Jones, we can experience the world like Indiana Jones.

7 **Go outside** – In nature, there are fewer of the cultural distractions that harm creativity and favour structured thinking (a concept that describes this kind of thinking wonderfully is mind-shaping culture). Grab a tent and find some wilderness.

***

Applying creative thinking in everyday life allows us to solve problems, but it may also promote our ability to learn. In chapter 8, we will get to know how higher-order thinking is involved in learning.

THE LEARNING BRAIN

If I asked you what comes to mind when you hear the word "learning", I am sure your answer would include schools and universities. You might also mention tutoring, webinars and self-learning for a profession or specialization. All these answers are correct, but they paint an incomplete picture because they relate to only one component of learning: acquiring linguistic knowledge.

Learning is far more than this. It includes the acquisition of skills, movements, sensations, emotions and much more. Learning encompasses all areas of life, too, and it is happening all the time, without us needing to go on a course. In terms of the brain, learning is a process of introducing long-term or permanent changes to the connections between neurons. Certain connections are severed or weakened while other ties are strengthened and new ones added. The brain itself is not happy about such changes; it is content with things as they are. To make the changes necessary for learning, we must convince it to promote our survival. This is where motivation comes into play, the virtues of which we discussed in chapter 3 but will now look at more closely.

Motivation

Learning without motivation is possible but any changes in the way the brain is organized will only be effective in the short term. This is evidenced by the following story, which a patient told me about her own experience.

From a very young age, she longed to play the guitar and, when she was five years old, her parents found her a music teacher. After only two or three lessons, she "lost the urge", as she put it, but her parents refused to accept this and forced her to continue playing. The forced lessons continued for five years, during which time she played in several concerts. It wasn't until the age of ten that she finally abandoned playing and her guitar gathered dust in the attic. More recently, she was handed a guitar at a family event and everyone begged her to play. "I took the guitar," she said, "placed my fingers on the strings, but I couldn't play. I had forgotten everything."

I wouldn't be surprised if you raised your eyebrows when you read the final part of our story. After all, the young woman had had lessons for five whole years as a child, and even played in concerts, so she must have had some musical skills. But let's not forget that all this was done because of the pressure from her parents. She had not wanted to continue, so she was not self-motivated to learn or to play the guitar. It is this last point that explains why the relevant changes in the brain's connections that had enabled her to play when she was younger were only temporary.

Think of motivation as a neural network that you activate for a given learning process, in which two forces act either separately or in unison. One such pair of forces is the pursuit of pleasure and the need to escape from a threat. Animal training (a form of learning) shows the effectiveness of activating these forces. A dog will be strongly motivated to learn tricks in its pursuit of pleasure – praise, a tummy rub or a treat. Animal training can also be achieved by using their natural desire to move away from a threat. For example, waving and shouting when herding animals into a pen.

Humans are not so different. Opportunities for promotion, bonuses and other benefits are effective because they activate one of our basic drives, which is the pursuit of pleasure. It creates the motivation to put more effort into our work, enrol on professional courses and improve generally. But the fear of redundancy, creating in us the need to avoid a threat, will have a similar end result.

The other force that can motivate us is meaning, and it is much more efficient than the opposing forces of pleasure and fear. Attributing meaning to a certain idea or object – whether that relates to your family, workplace or hobby – spurs the brain to view it as supremely important. The outcomes tend to match this level of personal investment and be significantly better than usual. In general, motivation that has its origins in meaning will tend to be stable, promoting physical and mental health and wellbeing, and awakening curiosity and creativity.

Different approaches to learning

In my lectures, I usually warn against having false expectations. "I can't teach you the topic of this lecture," I tell my listeners. "All I can do is provide you with the knowledge available, so you can understand it, and try to create a stimulus, so you will grasp its meaning. I present you with the raw material, then my job is done. The learning bit is yours and yours alone." That is because learning is a physiological process, affected by a number of factors. In this case, those factors are how motivated the students are to absorb the knowledge, their level of concentration during the lecture, the emotional state they are in, and whether they are well-nourished and have slept well.

Behavioural learning

The types of learning mentioned at the start of this chapter all involve the use of language, so are known as linguistic learning. But, as we will see later in this chapter, learning causes changes to occur in how things are organized in the brain and this can happen without using words, so it is possible to learn things just as well without depending on our linguistic ability. This is called behavioural learning because it is carried out using the motor, sensory and cognitive channels, and those used to control parts of the body, rather than language. For example, a person who wants to excel in a certain sport can learn techniques that will improve their performance without using language, as motor learning is required in this situation. The same goes for learning how to move a mouse to where you want on your computer screen or selecting fruit, as sight, touch and smell are most useful here. Countless other skills are integral to our daily lives. Although language can help with the learning process, it is not always vital. However, motivation is as indispensable for learning of the behavioural kind as it is for all other types of learning.

Let's jump back in time to the 18th century for an example of this in action. Imagine that you are on a sailing ship in the middle of the ocean. If you look up, you will most likely see a sailor standing at the top of the mast. He was not sent up there as a punishment, but because of his ability to sight land from far out at sea, which is not as easy as it might seem. If you or I took his place, we would not be anywhere near as good at it as him. The eyes and mind of that sailor would not be fundamentally different from ours, but he would have developed his ability to process visual messages to a highly sophisticated level by training himself to do this over time. In other words, by means of behavioural learning, he had developed a complex set of connections in the part of the brain that processes visual messages.

Returning to the present day, you may have come across wine-tasting courses. The goal of these courses is to enhance the experience of drinking wine by participants learning to improve their senses of smell and taste. This is not a false promise, as learning that involves the senses is behavioural learning. Now, let's say that a couple – we'll call them Sarah and Ben – see an advert for such a course. The subject appeals to Sarah and she decides that she'd like to sign up for it. Ben is not that interested but agrees to sign up too when Sarah asks him to go with her.

Now imagine that you meet Ben and Sarah two years later. You would probably discover that a new world of tastes has opened up for Sarah. Talking with her, she comes across as a real connoisseur, mentioning acidity, astringency, bouquet and vintages. It's not because she's out to impress you but because she has real enthusiasm for the subject and wants to share her knowledge. However, Ben's abilities have improved only slightly.

These different outcomes are to be expected as they started off with very different approaches to the subject. Sarah came to the course with a strong motivation stemming from attributing meaning to the subject as she was genuinely interested in it. Therefore, as her brain processed the sensory messages she received during the course, the information was maintained as long-term learning. However, the motivation areas of Ben's brain were not activated, even though he enjoyed the course. As a result, his learning was superficial and so was short term.

For a long time, it was widely believed that rationality was the dominant influence shaping our daily lives. As we saw in chapter 3, the findings of research have made it clear that this is not the case and, instead, emotion has the biggest influence on us.[1] As a result, brain researchers now agree that changes in our behaviour originate from changes to our emotions.[2] So, to ensure that we exhibit a particular behaviour, the brain must first change the emotion associated with it or regulate its intensity. The findings of an experiment called the marshmallow test, conducted by psychologist Walter Mischel at Stanford University in California, illustrate how emotions can lead to specific behaviours.[3]

In the experiment, the children participating ranged in age from three-and-a-half to five-and-a-half years old. Each child, in turn, entered a room where an experimenter was waiting, and sat down at a table on which one marshmallow had been placed. The experimenter then told the child that they had to leave and, while away, the child could eat the marshmallow. But if they held on until the experimenter returned, they would get another marshmallow. Two-thirds of the children gobbled up the marshmallow before the experimenter came back.

The children who managed to exhibit restraint were recorded on camera smelling, touching and staring at the marshmallow in the process of resisting the temptation to eat it. Similar experiments have been conducted elsewhere with comparable results.

Nonetheless, that was not the end of the matter. The researchers continued to keep in touch with all the children who participated in the experiment and made some fascinating discoveries.[4] Those who had controlled their emotions and restrained themselves from eating the marshmallow grew up to be adults who were more successful in their lives in various respects and could cope better with frustrations and pressures than average. It was concluded that this result was because they were blessed with naturally higher than average levels of emotional intelligence.

This does not mean that the fate of those who ate the marshmallow is inevitable. As we know from chapter 3, the brain is flexible, so we can enable children to develop their ability to manage impulses and emotions, encouraging them to become better at making decisions. It is helpful that the ability to differentiate between essential and non-essential is also part of the normal functioning of the brain. Hence, behavioural learning can be an outcome of emotional learning.

Biofeedback

The power of behavioural learning is that it gives us all kinds of abilities which are essential to our existence. Among other things, we can employ this type of learning to affect our pulse and blood pressure using biofeedback. For this, electronic monitoring is used. For example, if you were looking at the display of a device that was measuring your blood pressure, and you saw that your systolic blood pressure was 160, you would then concentrate on the idea that your blood pressure was lower. Most people then find that the figure on the display starts to go down.

The conclusion we can draw from this is that the power of concentration allows us to influence areas of the brain that control the body. However, it is not as simple as that. Once you are no longer looking at the display, your blood pressure will return to its previous level. This is because, at this stage, the necessary change has not yet occurred in the relevant neural network to maintain the behaviour as long-term learning. To achieve this, we must persist in the practice of concentrating on lowering our blood pressure or whatever else we wish to learn. After some time, the learning will become established and then you will be able to control your blood pressure without the help of biofeedback and later the brain will even do this automatically.

Learning and Alzheimer's disease

Western society attaches the utmost importance to both language and the acquisition of knowledge. The resulting huge emphasis on language-based learning has meant that we are neglecting a growing group of people whose ability to use language is impaired. This primarily concerns those with Alzheimer's disease and other types of dementia.

The cognitive decline that occurs in the early and middle stages of Alzheimer's disease causes problems for people in all aspects of normal functioning related to language, from thinking and memory to expressing themselves clearly and gaining new knowledge. At the same time, this cognitive decline does not affect their learning related to any of the other brain functions. Therefore, people with Alzheimer's disease can still employ behavioural learning to pick up new movements, improve the functioning of their senses and learn things related to emotions and control over their bodies. Yet, because their ability to use language effectively is diminished, they can be written off and it can be forgotten that there are other ways to communicate with them, teach them and improve the quality of their lives.

Learning knowledge

Similar to all the types of learning already mentioned, learning knowledge from sources such as books, lectures, movies, conversations and so on is simply a reorganization of the brain to enable the information to be retained in the long term. The only difference is the location where this learning is processed, which is the area of the brain that is focused on language. Also like the other types of learning, strong motivation is essential. The system of awarding marks or grades, familiar to us all from school, is designed to motivate us by means of two familiar forces: the pursuit of pleasure (good grades) and escape from threat (bad grades). As we know from our personal experience, this method works quite well. To enhance this, if the motivation comes from a higher-order place – from the areas of the brain associated with meaning – the learning will be embedded even deeper. This is what is meant by the terms deep or meaningful learning.

Philosopher and educator Benjamin Samuel Bloom originally established six levels for the learning of knowledge.[5] We will look at three of his categories that have been commonly used for the stages of learning, ranging from not knowing something in depth to knowing something very well. The accompanying descriptions note the scope of changes that take place in the organization of the brain for each type of learning and how the learning can be achieved. The goal is to go through these and reach the deep learning stage.

- **Superficial or shallow learning** – At the bottom of the pyramid, the changes made in the brain are minimal and expressed as a memory of the oral information (oral learning) acquired by repeatedly reading or hearing the information.
- **Intermediate learning** – In the middle of the pyramid, the changes made in the brain are more pronounced than for shallow learning, so the extra outcome is an understanding of the verbal information. The way to acquire intermediate learning is to reflect on important points or topics, divide the material into sections as subtopics and compare them.
- **Deep learning** – At the top of the pyramid. By this point, we experience a fundamental change in the brain and new connections are made throughout. This results in the internalization of the information and the way it relates to life events – our own and those of others. The formation of deep learning comes about via research into each of the subtopics as they appear in the material, thinking about how each one engages you and how this might be related to you or someone you know.

The level at which a specific piece of learning will take place depends on two components that enable knowledge to be learned: memory and thinking. For the purpose of illustrating the roles of these two components, let me compare the learning of knowledge to baking a delicious chocolate cake.

The raw materials are the foundation. Without good raw materials, we cannot reasonably expect to have a tasty cake at the end. Equally, if we do not use the right processes, however good the ingredients are, the result could be a useless ball of goo. So, continuing with this analogy, the raw material needed to learn knowledge is memory, and the process used to work on our memory is thinking. As with our cake, without the good raw material of a good memory and the right process of good thinking, we will not be able to achieve the fluffy and tasty cake of proper learning.

Memorizing material (learning by heart) is an accepted and commonly used method that we use to study a subject, mainly for exams. As I am sure you will have memorized facts, quotes and so on, I suspect you will raise an eyebrow as I tell you now that this type of learning is superficial learning. The reason for this is that when we memorize something, a memory message is sent to the neural network relating to memory and only that area of the brain is activated – the thinking component is out of the loop. There is nothing wrong with memorizing as one of the stages of learning but, in the absence of thinking, it won't get us very far if we want to be able to recall the information in the long term.

As its name suggests, intermediate learning goes deeper than the superficial kind. This is because it includes the extra ingredient of understanding the material being studied. For example, take the previous chapter in this book. On the assumption that you invested some time in thinking while you were reading it and properly understood what was described in the chapter, its content should now be etched on your memory. However, understanding only guarantees a moderate level of learning – something that I can attest to from my experience in my first year as a medical student.

I remembered what was said in the lectures and understood the content, but my exam results were average. The reason for this became clear to me later, when I found out that understanding only guarantees a moderate degree of learning. This happens because studying that doesn't go beyond listening to or reading the material we want to remember can't deliver all the goods. It is also why, when the exam questions referred directly to the content of the lectures, I had no problem answering them well. However, when the questions asked things such as what treatment is required for a sick child who comes in and has a particular diagnosis, to answer them properly, I would have had to have studied information that was not mentioned in the lectures. That extra step is part of deep learning.

While intermediate learning is achieved by fulfilling the memory and understanding requirements, deep learning involves also activating the higher-order forms of thinking, such as drawing conclusions, critical thought and creative thinking. As mentioned in chapter 7, critical thinking requires us to compare a specific text with other pieces of information previously stored in our memory, thereby enabling us to draw conclusions. Creative thinking involves processing information in ways that are multidimensional and multidirectional.

You may also recall from chapter 7 that creative thinking has to overcome our mental blocks as the brain does not want us to be creative.

It desires only one thing, our survival, so it is hooked on maximum efficiency and prone to processing data in a stereotypical way. That is why we label people according to character traits: it saves our brain some time, it is efficient.

Creative thinking is that which is outside the box of stereotypical categorization. Though it is now a common concept, the old Jewish Talmudic sages would not have come across it at all, yet they applied it around 2,000 years ago when they wanted to process raw material in a way that would achieve deep learning. I am referring to the idea of "quibbling", which is similar to the Socratic method and involves not focusing on the essence of something but seeking to reach a conclusion by means of extensive arguments about small and unimportant details, also considering biased opinions and even absurdities. A brilliant upgrade to this concept is learning in pairs, which is still practised to this day in Torah study. It is not unlike brainstorming – an activity now used extensively in all walks of life to help find creative ideas and solutions.

The success of all the variations on the brainstorming method rests on the fulfilment of two conditions. The first is that all participants in the discussion stay on an equal footing (no one gets to be the boss) because our brains intuitively adapt to align with the person perceived as having the highest status. The second condition is that equal weight is given to the opinions and ideas of all the participants; nobody should be criticized as talking nonsense.

Many studies that involved thousands of people, have looked at how we can enhance deep learning, and the findings provide us with some useful advice. One helpful tip is this: ask yourself questions. Each question forces you to zoom in on a certain component of the material you're studying, and this boosts the effectiveness of the learning process. Also, the mere action of posing a question induces slight mental tension, which sharpens the functioning of the neural networks relating to memory and thinking.

Further, as with many other phenomena related to the brain discussed in earlier chapters, deep learning has its roots in the lives led by our ancient ancestors. Their lives as hunter-gatherers were full of dangers. To improve their chances of survival and overcome the threats that came their way, tension created by the brain in response to a given threat was accompanied by a sharp increase in concentration and focus. This put them in a state of readiness to choose the right action and move quickly if required. Even today, a moderate amount of stress acts as a catalyst. So our memory and thinking improve and are clearer when, for example, we have a deadline. Nevertheless, it is a balancing act because excess stress inundates the system and has the opposite effect, disrupting learning.

Another piece of practical advice derived from the studies is that learning punctuated by breaks is much more effective than a longer period of concentrated study. If we need, for example, ten hours to prepare for an exam, we will achieve much better results if we avoid cramming our study into one or two days and, instead, hit the books for an hour a day. Learning is a physiological activity that creates new connections in the neural network. Breaks make it possible for this to happen in a more structured way than if the brain has to deal with large volumes of information at a time.

While these two methods have been demonstrated to improve your learning, there are two other approaches to learning.

One of these approaches is to read a text over and over again. However, repeated reading only serves the memory, so the best outcome is that we will be able to recite the text by heart. Doing this results in superficial learning, which does not generate understanding of the material. However, the ancient art of quibbling, Socratic discussions or brainstorming will greatly deepen learning.

The other approach is marking important sections in the text with a highlighter. The highlights will make us feel that we are thoroughly inside a text, but to learn from it, we must step out of it. There is nothing wrong with using a highlighter but we should only do this at an intermediate stage in the learning process.

Social learning

Even at the very beginning of the existence of human beings, the brain understood that we cannot fulfil all our needs without the help of others. We are social creatures because we had to be to survive. For this to happen, our brains needed to act in a similar way to those of the people around us, so we could work together. This led to the creation of a brain function that is unique to us and all creatures that have a relatively developed cortex. That function is social learning. This type of learning relies heavily on our ability to imitate. For example, we generally learn to use a tool by first seeing another person using it, and the same is true for many other activities that make up everyday life.

This ability to imitate is made possible by our mirror neurons. As their name suggests, these are nerve cells that ensure our brain activity synchronizes with that of the people around us. It happens automatically. If we see a person waving to us, the area of our brain associated with activating our muscles needed to wave back will be activated immediately but our hand will not always wave back. This is because another area of the brain comes into action, delaying the execution of the hand waving on the understanding that it has nothing to contribute to our survival if the other person is hostile, for example. At other times we may find ourselves automatically waving to a total stranger, such as when we are very tired. This is because fatigue weakens the brain's ability to maintain the delay. This mechanism explains the phenomenon of yawns being contagious. The more tired we are, the weaker the inhibitory action of the brain becomes.

Another example that illustrates how mirror neurons work is when people are waiting to cross the road. If one of them sees a gap in the traffic and sets off before the traffic lights turn green, then it often happens that some of the other people do the same. This is down to a momentary failure of the same delay mechanism to act, which usually operates because crossing the road is obviously life-threatening.

So far, we have looked at the synchronization of motor and sensory activities. Nonetheless, to thoroughly integrate into society – which is the brain's goal – adjustments also need to be made to everything related to our emotions. The brain activates the mirror neurons in this area, too, resulting in the general tendency of people to align their emotional state with that of the group. For example, if we see a loud argument between two drivers, we will feel angry as well, even though we were not involved in whatever caused the fight to start. Similarly, if we join a cheerful gathering, that general emotion will most likely have a positive effect on our own mood. However, if we join them when we have a good reason to be sad, that mood will cause the delay mechanism to prevent their good cheer from spreading to us.

Exposure to mirror neurons is only the first step toward social learning because what happens next depends on our motivation. Without motivation, the behaviours we are to imitate picked up by the mirror neurons will remain hidden in the brain and unused. The following story will help to explain this a little more.

A toddler at a kindergarten befriends a child with a stutter and, to his parents' dismay, he too begins to stutter. At the outset, it is only an imitation behaviour, initiated by the child's mirror neurons, so the situation is still reversible. But if motivation were to be added to the imitation behaviour, learning would occur and the toddler would continue to stutter.

The simplest way to prevent motivation developing is to ignore the stutter. If, instead, the parents were to plead with their son to stop stuttering, the opposite would happen because bringing the behaviour to the child's attention would lead to motivation developing. Another way this may happen is if, from the beginning, motivation were to become attached to the imitation. For example, if the toddler who stutters has charisma, is popular or receives preferential treatment from the teacher, then of course there will be deep learning that stuttering helps us to survive.

I have experienced a similar phenomenon myself. During my internship near the end of my medical studies, I was able to work for a short time alongside a senior doctor – one of the great experts in a field in which I had developed a special interest. He was an older man, an immigrant from Russia who spoke fluently, albeit with a slight Russian accent. After a few days, friends pointed out to me that I was pronouncing certain syllables differently, similar to the way in which the doctor pronounced them. Only a few days after the end of the internship did my speech return to normal. When I started working as a brain researcher, I came to understand why this had happened. My mirror neurons had got me to imitate the doctor's accent but, after we parted ways, the motivation to continue to do this faded, and so too did the foreign accent.

To take another example, the world of marketing makes extensive and effective use of mirror neurons, as this story of what happened to a friend of mine makes clear. He's a doctor of biology and a perfectly reasonable person. One day, he received a letter congratulating him on winning a widescreen TV and inviting him to come to collect his prize at a certain address. I met him the day after he went and he was very upset. "You won't believe it," he told me angrily. "I went to collect the prize and found myself signing up for a timeshare apartment, for life."

I wondered how this had happened and he told me that when he arrived at the address given in the letter, he found himself in a hall full of people. The organizers projected beautiful images of holiday apartments set in spectacular scenery. Many in the crowd were wildly enthusiastic and sufficiently impressed to crowd around the sales booth.

The sights and sounds, which were absorbed by my innocent friend's mirror neurons, found their way to the areas of his brain related to emotions, where they mixed with a deep motivation, one rooted in our evolution: to be part of the group. Only someone coming to the event with strong opposition to such a deal would manage to escape the brain's good intention of making us social creatures. My friend found out that most of those in the crowd were, in fact, employees of the timeshare company, planted in the audience to create an atmosphere of spontaneous enthusiasm. The mirror neurons obviously did not pick up on this trick, but it wasn't their job to do so.

To the brain, all types of learning are nothing but physiological processes that create new connections in the nervous system. The body needs proteins, sugars, fats, vitamins, minerals and other chemical substances to do this. Therefore, with all due respect to motivation and meaning, good long-term learning is not possible without proper maintenance of the hardware. Happily, the tools required for this are readily available: good sleep, proper nutrition, physical exercise and mental relaxation.

Some tips for improving your ability to learn

Here are some things you can do to promote deep learning of a new piece of text you need to remember when studying for a test or preparing for a presentation.

1 Repeated rereading of the text will only strengthen shallow learning of the text (learning it by heart), so don't be satisfied with that method alone.
2 Highlighting the important sentences in the text that you need to learn will only strengthen shallow learning to an intermediate level, but will not result in deep learning.
3 An excellent way to deepen learning is to learn in a group. After you have come to understand the text (reached an intermediate learning level), it is useful to meet up with one or more friends who are at the same learning level and together find out the links that exist between each of the ideas presented in the text and your everyday life.
4 Studies show that it is better to divide the learning into sections and not to study for hours in a row, as the periods when you are not studying help the biological connections between the neurons to occur.
5 Remember that the quality of the application of these and other methods to your learning will depend on the quality of the biological material that makes up your brain. Therefore, it is important that you sleep well, eat healthily and engage in social and physical activity in between periods of study.

***

Learning allows us to acquire new skills. The ability to learn linguistic information is essential for our personal development, and it involves memory and thought processes. The information processing that takes place in the brain changes the way our neural networks are organized. Specifically, it involves the making and breaking of connections between neurons. These changes are critical to our ability to apply our learning. Indeed, how well we learn can be characterized by the extent and depth of the changes made in the brain.

The information processing that happens during learning is linked to the way we perceive the world and behave in it. In chapter 9, we will come to understand this more as we get to know how the brain presents our personality.

CHAPTER 9
OUR RELATIONSHIP WITH THE WORLD SHAPES OUR PERSONALITY

I firmly agree with the general negative response of neuroscientists that the human brain is like a computer. It is clear beyond any doubt that our brain is much more complex, more sophisticated and infinitely more capable than any collection of semiconductors. Still, for the purpose of simplifying the subject of this chapter, I have allowed myself to use some basic concepts taken from the field of computing to explain how we humans interact with our environment. The first of these is to use an input–output model to look at how the brain processes information that comes into it via the senses.

Our relationship with the world happens in three stages, found both at the conscious level (in the cortex) and the unconscious level (in the subcortex), which are the:

- **inputs** flowing from the senses to the brain
- **processing** of those inputs
- **outputs** of behaviour

Here is an example of how this trio functions in everyday life.

Jada turns on her computer and notices that the cursor on her screen will not move. The visual input that reaches her brain is simple and to the point: the cursor is stuck! Her brain processes this input to create an understanding of the situation and how she feels about it. The output will be the drive to create behaviour that promotes survival, which either directly or indirectly is the brain's one and only goal. In this case, one of the options is to tap the mouse lightly. Another option is to check whether it's connected, the batteries are dead or it needs to be charged.

The inputs and processing stages

The eight senses continually provide us with inputs or stimuli (see chapter 6). The path these pieces of sensory information take is defined as "bottom up" (BU), and their first stop after initial processing by the sensory organs into messages is the subcortex. From there, they will usually continue to the cortex. However, some messages from the senses stop in the subcortex – the brain does not make us aware of them. Often this happens when messages are too weak or too short to continue on the BU path. An example of this is illustrated by an experiment in which participants were seated in front of computer monitors and shown various images of faces in short flashes.[1] When the participants were asked to say what facial expressions they had seen, they were unable to. That is because, although these details had been registered by the brain in the subcortex, the messages had not travelled on to the point on the BU path where the participants would have been aware of them. What we receive from such flashing sensory inputs are "subliminal messages".

Awareness is not a necessary condition for the brain to promote survival, as we have seen. Even creatures without a cortex, such as fish and reptiles, can escape from danger because the subcortex operates automatically to keep them safe, without the need for approval or any reaction. Messages coming from the sensory organs are simply recorded in the subcortex and undergo processing that takes things to the final stage, which is the output.

Even though we humans have a developed cortex, we saw in the experiment featuring subliminal messages that when inputs from the outside world reach the brain as a flash, we will receive and process them. However, we will only formulate outputs in the subcortex, while all this completely bypasses our awareness. An example of this happening in our everyday life would be a driver suddenly slamming on the brakes *before* consciously realizing that a child had run out in front of the car. A report from the brain about the child's carelessness and the automatic reaction of braking would enter the driver's awareness only in retrospect.

Such emergencies are one reason that automatic responses are triggered by the brain. Another reason sensory messages go no further than the subcortex is that the brain decides they are irrelevant to our survival. An illuminating example of this is situations in which a part of the body transmits a pain signal but we feel no pain. This happens because the pain does not threaten our survival. Even though the pain has been registered in the subcortex, we are not made aware of it, so we do not feel it. If, in such a situation, someone asked us, "Are you in pain?", we would honestly answer, "No."

One way in which this automatic response to pain has been used to advantage is to perform some major surgery (for example, involving opening up the stomach) under hypnosis.[2] There is no need for anaesthesia as the patient does not feel any pain. This is possible because the hypnotist conveys messages of calm, peace, vitality and wellbeing to the patient, who then enters a meditative state. Thus, the messages received by the brain are that all is well, so, as the brain's priority is to maximize survival, it deems the pain signals unnecessary and blocks them from going any further than the subcortex.

Alexander Solomonovich, who teaches hypnosis, told me about one hypnotist who, in the middle of a major operation, turned to the patient and said, "You are now in a calm and pleasant place. You feel no pain, but your right hand represents an area of the brain that sensations of pain can reach."

After saying this, he handed the patient a notepad and asked him to write down how that area of the brain felt. The patient wrote, without hesitation, "Damn you all, you're killing me", so the hypnotist restored pain control to the hand once more. Yet, all the while, the patient's calm facial expression remained unchanged. This proves that pain messages were being received by the brain; it simply was not passing them on, so they had not entered the patient's awareness.

As noted, the processing of sensory messages does also happen in the cortex. I'm sure you won't be surprised to learn that it uses completely different processes from those used in the subcortex.

The processing of messages in the subcortex

As the subcortex is the most ancient part of the brain, the way it processes messages is with a complete focus on meeting its sole clear goal: our survival and ability to reproduce to continue the human race. It is focused on factors such as finding shelter and food, maintaining the correct body temperature, dealing with enemies and predators, courtship and passing on our genes. To help it achieve its goal, the subcortex has two "inspectors" – one optimistic and the other pessimistic.

The optimistic inspector is put to work when the messages received indicate that survival and reproduction are guaranteed, such as it would in a male snake that had finished a hearty meal and saw a female snake nearby. But the pessimistic inspector will operate when all the messages are to the contrary, such as our snake being desperate for food and there are no female snakes around.

The optimistic inspector sends the messages it receives to be processed in a neural network in an area of the brain called the basal ganglia, directing it to create an emotion of pleasure. The messages received by the pessimistic inspector go for processing in the amygdala instead, where an emotion of threat is created.

In our daily lives, there is a non-stop flow of sometimes contradictory messages, so both inspectors are on duty all the time. Let's take an example from the daily routine of our ancient ancestors that shows how conflicting messages are dealt with. The hero in our example has just completed a successful hunt and returned home with enough food for the coming days. Therefore, the messages that reach his brain while he is busy butchering the animal are optimistic. However, some other less pleasant messages come to his attention presently, of his friends mocking him for what they saw as his timid behaviour during the hunt.

Now let's think about how such a scenario might play out in our modern age. After a lot of hard work, Tom manages to iron out a glitch that was slowing down a project crucial to the future of the company he works for. Cheerful and happy, he tells his immediate superior about this in an email and now he's waiting to hear a thank you and, maybe, news of a bonus.

The response hits his inbox a few minutes later: "Thank you for your efforts, but it has been decided to withdraw from the project as the company is being wound up."

Most of the time, we experience pleasure and threat simultaneously, so it is the relative intensity of the various messages that will ultimately determine our emotional state and how we will react. In this case, which inspector would have the upper hand and deal with the communication from Tom's manager? Whichever one proves dominant will determine whether the basic processing of this input will be to create an emotion of pleasure or threat, and determine the intensity of that emotion. This is a lower-order, unsophisticated type of processing, only for the here and now, but it expresses itself as an element of our personality. As mentioned earlier in this chapter, higher-order processing happens in the cortex.

The processing of messages in the cortex

The processing that happens in the cortex is affected, to a certain extent, by factors such as our knowledge, values and prevailing cultural and social norms.

The cortex, like the subcortex, seeks to promote our ability to survive and reproduce but, unlike the subcortex, which is concerned only with the here and now, the cortex takes a long-term view. This is sometimes at the expense of our immediate comfort. For example, if your manager teased you, this would be processed in the subcortex first and possibly in a way that would trigger the feeling of being threatened. Nonetheless, the cortex would then counter this feeling by concluding that what your manager said was in jest, but you would have experienced momentary discomfort before the logical, reasonable interpretation prevailed.

Here's another scenario. A man is on his way home and he's feeling hungry. The sensation of hunger that has been activated, when processed in the subcortex, results in the output that this is a threat to his survival. His walk takes him past a kiosk selling doughnuts. The sight and the aroma aggravate his hunger pangs and so the subcortex boosts the feeling that his survival is under threat. However, the output from processing in the cortex is to focus on the meal that he will have when he gets home, so the feeling of threat is turned into one of pleasure by the expectation of something pleasant.

The outputs stage

This stage is reached after the sensory inputs have been processed.

The outputs of the subcortex: responses of fear or pleasure

As we have seen, the outputs of the subcortex are designed to promote our survival in the moment, without consideration being given to any long-term aspects. Therefore, when the processing of a sensory input ends with the subcortex activating an alert of a threat, the output will be one of the three responses to threats discussed in chapter 3 – fight, flight or freeze. Which of these responses is the dominant one varies from person to person. Let's take an imaginary situation that illustrates this idea, now well-established by scientific research.[3]

Someone dresses up in a large bear costume and stands still in a store, then moves when people walk by them. Most people will run away in a state of panic, as an automatic response to the thought that somehow this is a real bear, while others will freeze in place and some will instinctively start to attack the bear. From the type of reaction people have in such a situation, neuroscientists can hypothesize how they will behave in threatening situations encountered in everyday life.

Another quite different response that can be the output when sensory messages are processed in the subcortex is pleasure. Pleasure shows that the brain understands that an existing situation promotes our survival, so the output will be behaviour intended to ensure the situation continues for as long as possible. We see this pattern in dog training, as I recently had the chance to witness.

I had been walking along, but then had to wait for the traffic lights to change and a man and his dog stopped beside me and waited too. The dog immediately sat down next to his owner. When the lights changed to green, we all crossed the road, but stopped again as the next lights were red. Once again, the man stood still and the dog obediently sat without being told to. I complimented the man on his dog's behaviour. "It's a very simple matter," the man explained. "When I was training him, I stroked his head every time he acted like that."

This approach worked because the stroking activated the pleasure zones in the dog's brain, and the repetition forged a link between the feeling of pleasure and the behaviour of settling down whenever his owner stood still, which soon shaped the dog's behaviour so it became a habitual one. The dog's response was designed by its brain to ensure the continuation of the pleasurable situation.

Deriving pleasure from an external source that does *not* help us to survive and can even be harmful, such as addiction to drugs, alcohol or sugar, is problematic and often dangerous. Such behaviour can seem counterintuitive but it results from uncontrollable impulses that cause people to behave in ways that are solely focused on ensuring the pleasure areas of their brain continue to be stimulated, with no thought given to positive motivation or outcomes. Thus, people addicted to drugs will not hesitate to steal and commit other crimes to get the money for their next fix.

The outputs of the cortex: behaviour and personality

So far, we have looked at the outputs of the subcortex. As mentioned, there are differences between its outputs and those of the cortex. The cortex has the ability to plan behaviour. All animals blessed with a cortex can do this, but we humans have the most highly developed cortex of all, with by far the best capabilities.

After messages have been processed in the cortex, with ensuring our long-term survival and the chance to reproduce as guiding principles, outputs are created in the form of impulses to behave in ways that will achieve these goals. The behavioural outputs are defined as planned when compared with those created by the subcortex because the cortex takes into account social and environmental elements. For example, a considerate driver entering a car park and finding it full will not have the urge to park in a space for users with disabilities. This is because the cortex will view this as both immoral and risking a fine, so it would not create an output influencing the driver to park in a disability space. However, the subcortex would.

Let's consider another example. Jenny goes over to her friend's and is invited to stay and have dinner with the family. She sits down at the table with them and sees that one of the children has not sat down yet so, even though she's hungry, she does not dig in until he arrives. During the meal, she also exhibits good table manners.

Now, let's compare the outputs of the cortex we've seen played out here with what the outputs of the subcortex would be in the same situations. As noted, the driver would park in the disability spot. The problem here is that, even if this would have solved the immediate problem, the hostile reactions of passers-by, especially if any of them knew the driver, and the fine would cause them harm in the long term. The same is true of our dinner guest. If Jenny had satisfied her hunger straight away and not shown consideration for those around her, she would have solved one problem sooner, but would probably not be invited again, so this would have caused long-term harm to her friendship with the family. Therefore, the cortex, by including social and environmental norms in its processing, increases the likelihood that we will achieve its goal of increasing our chances of being able to survive and reproduce.

There is a further element that influences how people act and this can have a huge effect on the course of their life. Consider an evil murderer and a law-abiding citizen. They have the same brain processes:

inputs of information are received, processed and the resulting outputs determine their behaviour. So how come they are so different?

The answer lies in the fact that each and every person has a unique personality. This means that each person processes the messages they receive in a unique way, which creates unique behaviour. The different aspects of the uniqueness of our processing and behaviour are defined as character traits. It is customary to characterize these features along two intersecting axes.

1 introvert–extrovert
2 stable–unstable

The introvert–extrovert axis represents how individuals usually express their emotions. Thus, the behaviour of someone who is an introvert, whether they feel threatened or at ease, will always be restrained. This happens naturally, without them trying to control their behaviour.

An extrovert will, instead, let all their feelings show in the way they behave. For example, when they feel threatened, they will shout out loud, run away or attack what is threatening them.

In the context of the stable–unstable axis, "stability" means that similar inputs in similar circumstances will result in similar, expected behaviour. For a stable person, so long as the general situation does not change drastically, similar inputs will produce pretty much the same feelings and reactions.

For a person who is emotionally unstable, the situation is very different. It is impossible to predict their emotions and behaviour, even in response to ordinary events.

To see what behaviour might be expected from people at the two ends of this axis, consider the reactions of two people to a long hoped-for declaration of love. A stable person will respond to such words with joy, without any memories of previous romantic disappointments clouding her feelings. However, for an unstable person, the same situation could provoke one of several different reactions, either happy or sad.

In the general population, each of us stands in a different place on both the introvert–extrovert and the stable–unstable axes. Someone who is located at the extremes of both axes, in one combination or another – extrovert–unstable, introvert–unstable, introvert–stable or extrovert–stable – is said to have a rigid personality. This may manifest in a clinical personality disorder. If someone is located more centrally on both axes – and even able to move easily along the scales – they will be defined as having a flexible and healthy personality.

An example may be helpful here. Imagine that Max, who has a highly extrovert and unstable personality, is at a job interview. The interviewer throws him by asking, "What would your former manager say about you?" Sensing this as a threat, Max's extrovert side might cause him to react bluntly and his instability might mean that things could deteriorate, to the point where verbal or even physical violence occurs. The degree of threat perceived by someone in this situation with a more balanced personality would be at a lower level, so their reaction would be more moderate. They might look grumpily at the interviewer, shift uncomfortably in their chair and shuffle their feet, but they would not lose their self-control.

Therefore, it seems clear that a healthy personality results when someone's brain responds in a way that places them – at any given moment and according to circumstances – in the correct place on both axes. That is, they naturally behave differently and in step with the various situations in which they find themselves. For example, they can be tough and inflexible when negotiations are required, but mild and tender when someone is experiencing some difficulty or when interacting with family. Unfortunately, not many people are endowed with such balance – generally, we easily blow a fuse when it is totally unwarranted.

From this, we can see that character traits influence behaviour, but our discussion about human behaviour doesn't end here. Let's move on to two types of behaviour that characterize us. The first is personality bypassing behaviour, also known as planned behaviour.

For example, if someone knows that they have to be at their best in a job interview, even though their nature is generally extrovert and unstable, they will do everything they can to keep this hidden. The good news is that each of us can engage in planned behaviour, up to a certain point. This skill of behaving in a way that bypasses our personality lets many of us act properly in uncomfortable social situations.

The second form of behaviour is reactive behaviour, which appears when something takes us by surprise, and is an honest reflection of our personality traits. To illustrate this, let's drop in again on Max, the job seeker, who is now midway through the interview. So far, apart from that question, everything has gone well in terms of his planned behaviour. But then, as a tactic to draw out his true character, the interviewer tells him, "Yesterday, someone you worked with at your last place called to warn me about you."

As Max wasn't expecting this turn of events, his behaviour would naturally be reactive, faithfully reflecting his character and personality. If he was to get all riled up, jump out of his chair and shout, "Who told you such a vile lie?", this would show that he is an unstable, extrovert and rigid person.

The response of a person with a balanced and easy-going personality would, instead, likely express curiosity and say something like, "I'm very surprised . . . I can't imagine why they would say that . . . can you tell me the name of this person?"

In psychiatry, traditionally, several types of personality disorder are defined by their abnormal rigidity and extreme, unrealistic feelings of being threatened. Here are some of the most prominent of these.

- **Schizoid personality disorder** – People with this disorder feel threatened by any social contact. All other interactions, whether with pets, plants or computers, present no problem and are even desirable. However, people with this disorder will have a nervous breakdown if they are forced to be in the company of people – something that may, in fact, lead to a psychotic seizure.
- **Paranoid personality disorder** – Individuals experiencing this tend to have a long history of distrust, attributing malicious or harmful intentions to other people in any situation they deem unsatisfactory. For example, if a server doesn't hurry to take their order, they will be sure that they've been ignored because of their ethnicity, gender or some other such reason, not simply because the restaurant is busy. This feeling will be expressed in uncontrollable anger, taken out on the hapless server.
- **Narcissistic personality disorder** – It's widely misconstrued that people with this disorder have an excessive sense of self-importance whereas they suffer, first and foremost, from a complete lack of self-worth. They try to overcome this by seeking appreciation and respect from those around them, but it's never enough – they are like a bottomless pit for praise. If they feel unappreciated, their response of feeling threatened will overcome them and they may reach a point where their emotional functions collapse. This may manifest as deep and prolonged depression or uncontrollable rage. To gain sympathy and appreciation, they will make every effort to always be the centre of attention.
- **Borderline personality disorder** – This condition is more common in women than men and expresses itself as sharp and unexpected transitions from feeling threatened to being overcome with pleasure. This chaotic situation may cause people with this disorder to engage in an excessive

search for thrills in areas such as sexual activity, alcohol or drugs, and cause extreme instability as they go about their daily lives. It follows that those with this disorder usually have trouble maintaining relationships. When the perceived level of threat increases, this is accompanied by real suffering that causes mental anguish. In extreme cases, people with the disorder seek to convert abstract mental pain into focused physical pain by self-harming.

- **Antisocial personality disorder** – This is more common in men than women and is often seen in prison inmates. It manifests as what I describe as the sanctification of the ego. Thus, an injury, even a small one, to the dignity of a person with this disorder will trigger feelings of being extremely threatened, so they will react impulsively, without any consideration for social conventions. For example, if during an argument about a parking spot, one person was to swear, someone with this disorder will not hesitate to respond by pulling out a knife and stabbing them.
- **Dependent personality disorder** – Those with this disorder are unable to make their own way successfully through life; they need help with making day-to-day decisions from someone they consider to be an authority. When they find themselves cut off from their source of authority, they perceive it as an existential threat, which might manifest as clinical anxiety or severe depression.
- **Obsessive-compulsive disorder (OCD)** – The best way to describe people with this disorder is to say that they are extreme perfectionists. They do not recognize the possibility of compromise. Things can be only perfect or flawed, and between these two extremes there is an absolute void. If the expression of the disorder is cleaning their home, you might find the OCD person mopping and vacuuming the floor, cleaning the kitchen, bathroom and windows or dusting several times a day. However, due to their pursuit of perfection, many of those with OCD are high achievers in various areas of life, but this often comes at a high price.

There are no miracle cures for any of these disorders. Nonetheless, the knowledge that the brain is flexible should enable techniques to be developed that will help people affected by them to regulate their emotions and exert self-control. The necessary condition for the success of these techniques is a strong motivation to change, but this is neither easy nor simple for them.

***

Understanding the brain code allows us to fine-tune the subtext of our interactions with people. Having an accurate and appropriate subtext helps us to make a good impression and generates appreciation and sympathy.

As you go about your daily routine, remember to smile more at the people you meet, move gently and try to moderate the volume of your voice during conversations. The next day, ask the people you encountered how they felt about meeting you.

CHAPTER 10

TAKING CARE OF OUR GREY MATTER

Not so long ago, it was generally thought that if our brain was "programmed" correctly, it would be guaranteed to function normally. Neuroscientists believed that to maintain the integrity of our memory it is enough for our "memory software" to be in good order. Only at the beginning of the 21st century did researchers come to accept that even the best software will only run properly on hardware that is properly looked after. The "hardware" in this case is the billions of cells that form the brain. Each one of these cells is made by the body from proteins, sugars, fats, vitamins, minerals and water. Not only is the rest of the human body also made from these raw materials but so are all living creatures on Earth.

The definition of a living being – an organism – is that it is a structure that can survive and reproduce. This vast category includes those that consist, at the most basic level, of a single cell, such as amoebas and bacteria, as well as the more complex plants, fungi and corals, right up to the animal world, which ranges from fish to reptiles and birds, and on up to humans. The cells of all these organisms are made from the same raw materials. During the life of every living being, there is natural wear and tear of the cells with proteins wearing down the quickest. The food that all living things need should continually provide fresh raw materials, which are needed to deal with this wear and tear. Conveniently, as all living things are made from the same raw materials, plus water and minerals, we look to plants, animals, fungi and bacteria to supply what we need.

A deficiency in one of the components needed by our cells is enough to degrade how well our brain functions. For example, omega-3 fatty acids are essential for ensuring the healthy functioning of our memory. An omega-3 deficiency will have an effect on all brain activities, such as our sense of smell, cognitive abilities, including mood and thinking, and even movement of the muscles and our physical health. (This comes about as a result of damage to the functioning of the immune system, blood pressure, sugar regulation and more.) Therefore, it is clear that proper nutrition is a precondition of the brain to function normally. Yet nutrition alone is not enough; we also need to take care of all parts of our body by getting regular exercise and setting up good sleep habits.

How we move

Before we look at how exercise helps us, let's take a step back and think about how the brain activates our muscles. The brain can contract our skeletal muscles by using either the voluntary or involuntary system. The voluntary system originates in the cortex. An example of a movement that it would activate is intentionally waving your hand.

The involuntary system, which originates in the subcortex, is activated when the brain realizes that a quick, automatic response is required to help us survive. This might be repositioning the body so we don't fall or moving quickly away from danger.

The movement of the skeleton in everyday life happens as a result of these two systems operating together. For example, riding a bicycle is a voluntary action, but contracting the right muscles to keep you balanced while you ride is an involuntary type of movement. Moreover, this operation continually changes, depending on the contours of the ground you're riding on and your speed.

The area of the cortex responsible for voluntary movements consists of many millions of nerve cells, with very long axons – up to 1m (3 feet) long or more. These axons group together into a kind of cable, which is part of the spinal cord. The messages from each branch of the nerves reach the relevant skeletal muscles as electrical signals, and it is these signals that activate the voluntary movements. For example, if you want to send a text on your phone, the area in the cortex relating to muscle movements (called motor skills) will be activated, and it will send electrical signals to the relevant fingers to perform the operation.

Adjacent to the area in the cortex responsible for such motor skills is the part that serves as our motor memory. There, the details of how we perform all manner of motor skills are stored, from brushing our teeth and combing our hair to tying our shoelaces or playing a certain piece of music on the piano. When we learn a motor skill and the brain understands that it is important for our survival, it gives us great pleasure and creates in the same area a memory of how to do it. This memory will include the order in which the muscles need to be activated, the strength and pressure that need to be applied and so on. When the memory has been consolidated (by links forming between nerve cells), performing that skill again in the future will be automatic.

All these voluntary movements are subservient to other extensive areas of the brain, including those responsible for mood, motivation, concentration and self-confidence. That's why, even though we have, for example, acquired great skill in playing that piece of music on the piano (because our brain has created a motor memory for it), we may not perform it too well if we lack self-confidence when we go on stage. This understanding that has come to us from brain research has implications for many areas of everyday life.[1] Among other things, the success of physical rehabilitation after an accident or stroke largely depends on the mood of the person, their self-confidence and levels of motivation. The more positive these aspects are, the faster the recovery process and the better the results. The same is true when training for various sports.

The area in the subcortex responsible for *involuntary* motor actions is the cerebellum – a name derived from the Latin for "little brain". Like the area of the cortex responsible for voluntary movements, the cerebellum is also composed of many millions of nerve cells. Their axons transmit messages that cause the skeletal muscles to contract or release. In the case of the cerebellum, these messages are all about involuntary movements. These can be ones that enable us to balance, controlling movements that need to be made simultaneously for the purposes of coordination, or fine motor skills. These are precise movements for a specific need, such as holding a pen to write.

Exercise

Exercise or physical activity is the movement of the skeleton for the purpose of strengthening the body, and it is achieved by contracting and relaxing the skeletal muscles. This can be likened to playing with a puppet. To make its arm move, you have to manipulate the string connected to that limb. Instead of strings, we have tendons, which connect our muscles to bones. When a muscle contracts, the tendon stretches and this produces the required movement of the bone. Activating our muscles to engage in physical activity is a prerequisite for a healthy, fully functioning brain.

Physical activity can be divided into five main categories.

- **Aerobic exercise** – Activities in which a variety of skeletal muscles work at low to moderate intensity over time, such as walking, jogging, cycling or slow swimming.
- **Strength training** – In this form of exercise, the relevant muscles are worked at high intensity, so the activities are short and limited, such as lifting weights or training using equipment at a gym.
- **Balancing** – To stay upright, the cerebellum must initiate muscle contractions in specific areas of the body. The big challenge for the cerebellum is when we stand on one leg and shift our weight or swing our limbs from side to side. In such a situation, the cerebellum instructs the body to contract multiple muscle groups to maintain our balance.
- **Coordination** – Try to rotate your thumb while folding your fingers on the same hand down one after the other. Can you do this? To be honest, I couldn't, but it involves performing at least two different movements at the same time. In our daily lives, we often do this sort of thing. When we drive, our hands turn the steering wheel, while our feet move between the brake and accelerator pedals as needed.
- **Fine motor skills** – As part of the process of maintaining our balance, the cerebellum continually receives reports from the limbs about their positions in space. These reports are also used when we employ our fine motor skills, such as threading a needle or putting a key in a lock. These seemingly different activities have two elements in common. One part is stationary and the other is moving. The cerebellum knows where the moving part is and gently directs it to its stationary target before we are aware of it. This is because having awareness of the location of the needle or lock will slow down the completion of these tasks because the information processing takes place in the cortex, which works at a slower pace than the subcortex.

All these types of physical activity help to maintain the hardware of the brain, but how do they do this?

In the case of aerobic exercise and strength training, it's quite simple. During these activities, the muscles involved consume large amounts of oxygen and other essential chemicals – substances that are carried in the bloodstream to all parts of the body. Now it happens that the increased supply of these substances reaches not only the muscles being used but all parts of the body. Supplying them separately would not be practical, so the whole body enjoys a boost. This means that the brain benefits from this generosity, and this makes the essential maintenance of its neural networks possible.

It is not surprising that there has been an increase in the numbers of people who extol the wonders of hyperbaric pressure chambers. These are used in medical situations and for divers, and also for therapeutic sessions at clinics, as they deliver an increased supply of oxygen to the body, including the brain.[2] The positive results of these treatments are indisputable, with people reporting a marked improvement in brain structure and function.[3] However, balanced aerobic exercise can, in my opinion, achieve the same goals, and at a far lower cost. The exception to this advice would be if someone has limited mobility. Then, pressure chambers are an effective means of improving brain function.

The other physical activities listed that involve balancing, coordination and fine motor skills also help to develop hardware in the brain. Evidence for this has been provided by the work of neurobiologist Rita Levi-Montalcini, who was a joint winner of the Nobel Prize in 1986 for her breakthrough in understanding the idea of the flexible brain.[4] Levi-Montalcini discovered that meaningful changes in brain structure will only occur if the nerve cells are in an active state. Moreover, she proved that if the brain itself is not active, the connections between the nerve cells will start to unravel, resulting in a decline in the brain's cognitive and motor abilities, as well as those related to physical health and control of the body's organs.

For our nerve cells to be active, we need to lead an active lifestyle, which includes not having routines. Repeating actions creates patterns of routines and these quickly become functions that the brain does automatically, which don't require it to make an effort. However, when the brain is exposed to a new and unfamiliar activity, large areas of the brain are activated to implement the necessary input-processing-output process. Aerobic exercise and strength training involve activities that are monotonous, so their contribution to improving the brain's software is limited. Physical activities requiring balancing, coordination and fine motor skills are much better as they are varied and they can surprise and challenge the brain. Indeed, contemporary brain research attaches great importance to an active lifestyle, noting that it is a condition for successfully engaging the abilities of the flexible brain.[5]

Sleep

It was common at one time to think that we sleep to give our brain a rest. Only in more recent decades have brain researchers discovered that this concept is fundamentally wrong.[6] Now we understand that during sleep the brain is not resting but very active. This activity is essential for the maintenance of our grey matter. All animals must sleep to maintain their brains, although there are some major differences in how they accomplish this. Some birds sleep while flying, fish must sleep while swimming, horses sleep while standing, but we humans like to be horizontal.

The following example illustrates the huge importance of sleep. A person ignores the feeling of tiredness creeping over them as they drive along a motorway. Their fatigue increases until they fall asleep at the wheel. At first glance, this does not make sense because the brain's paramount concern is the driver's survival and it's clear that this is in jeopardy when they fall asleep while driving. But the brain has no choice in the matter because adequate sleep is essential if it is to process information flowing to it from the senses.

Although the brain knows that falling asleep at the wheel may be fatal, it is unable to prevent this from happening because the neural networks are overloaded with sensory stimuli that have not yet been processed. Experience shows that, in such a situation, approximately ten minutes of sleep is enough to clear the space needed for the backlog of sensory information to be processed, so if the driver pulls over and has a short rest, all should be well. This is a bit like noticing that your phone battery is almost out of power when you need to make an urgent call. Charging it for a few minutes will give you enough power to make your call and then you can charge it fully later.

The results of brain research have not yet given us an unequivocal explanation of how exactly sleep works to improve our brains. But studies examining brain activity in cases of sleep deprivation have proved that sleep is vitally important for everything related to how well the brain functions. In one study of rats, which were deprived of sleep for a long time, the researchers saw a continuous decline in the general functioning of the brain.[7] It affected their motor, sensory and cognitive functions, and control over their organs. The experiment ended, sadly, with the death of all the rats (note: this experiment was carried out before legislation to protect vertebrate animals, the European Convention for the Protection of Vertebrate Animals used for Experimental and other Scientific Purposes, came into force in 1991).

Of course, no such extreme experiments have been carried out with humans. However, studies have been conducted in which researchers followed humans whose sleep was abnormal. They noticed an increase in the incidences of medical conditions such as malignant tumours and heart disease, and in mental health conditions such as memory loss, impaired concentration, depression and anxiety.[8] The studies that focused on long-term night shift workers, such as nurses, security guards and flight attendants, provided clear proof of a relationship between sleep and health.[9] This does not necessarily mean that everyone working night shifts will suffer from the physical and mental health conditions that have been mentioned, but the data indicate a connection between lack of sleep and an increase in these symptoms.

Clearer evidence of the importance of sleep is provided by people who develop fatal familial insomnia, a rare and deadly genetic disorder, first identified in 1986. As its name indicates, it is hereditary, but is known to afflict only around three dozen families in the world whose offspring have the gene mutation. Sadly, people who have fatal familial insomnia lose the ability to fall asleep and die slowly from exhaustion and the consequent systemic collapse of the body's functions.

Sleep can be defined as a temporary and cyclical physiological state that is essential for the brain to function, and during which there is a decrease in the stimuli and messages being sent to it. The direct connection between sleep and brain activity is no longer in doubt, but brain research has still not managed to provide a complete picture of the mechanism that makes it happen. However, by systematically monitoring sleep patterns of many people, some prominent characteristics of normal and irregular sleep have been identified.

Normal sleep

One notable characteristic of normal sleep is that it consists of up to four or five phases. Each sleep phase lasts about 100 minutes and consists of several stages. The final stage is characterized by rapid horizontal movements of the eyeballs. If you close your eyes and repeatedly look left and right without moving your head, this will simulate what your eyes do when you are asleep. This final stage is called rapid eye movement (REM) sleep, because of what our eyes do. It lasts about 20 minutes and this is when we dream.

Another biological characteristic of normal sleep that reinforces our understanding that sleep is cyclical, is that during sleep there's a regular pattern of electrical activity in the cortex. Tracking the electrical frequencies in the brain has revealed that during each sleep cycle there is a fixed pattern of electrical activity and any disruption of this indicates when there have been disturbances. During the phase of REM sleep, when we have the dreams we later remember, the brain enters a state of vigorous activity, as though we were awake. Some parts of the cortex are even *more* active during REM sleep than when we are fully conscious.

Brain research has discovered some intriguing insights into the relationship between dreaming and brain health. For example, to manage information that has reached the brain during waking hours the cortex needs to work overtime, and because our awareness is activated to process this information, dreams are created. There is no contradiction between this idea and the psychological theories that claim dreams fill an important role in bringing conflicts to our awareness.

So that sleep can perform its role in maintaining the brain, two other conditions must be met: the timing of our sleep and its duration. Regarding timing, it is customary to divide the animal kingdom into two basic categories:

1 **owls**, who are nocturnal, sleeping during the day and awake at night
2 **larks**, who are diurnal, sleeping at night and awake during the day

Humans are, by nature, larks and it is important that our sleep takes place at night. Only during the hours of darkness can certain physiological processes take place in the brain and other organs. These are essential if sleep is to be effective. One such process is the secretion from the brain into the bloodstream of the hormone melatonin. One of the functions of this hormone is that it protects the deoxyribonucleic acid (DNA) in cell nuclei. Therefore, the sleep of a person who comes off a night shift and goes to bed in the morning will only go partway to restoring their brain, even if they sleep the necessary number of hours and wake up feeling refreshed. They are like a hungry person who has a snack and gobbles it down. The result is that they might feel full, but the snack's contribution to their overall nutrition will be minimal because its nutritional value is low. The snack is junk food, and day sleep instead of night sleep is junk sleep. From this, it should be clear that the notion of "catching up on some sleep" does not work when it comes to looking after our brains.

The other condition is that we need to have the right amount of sleep for it to be truly restorative. Opinions as to how much sleep we need have shifted. Not so long ago, it was believed that six consecutive hours of sleep was ideal. It has now become clear that this is not enough and the view is that we need eight hours of sleep.[10] It also turns out that too much sleep is just as harmful as too little.[11] These findings are consistent with the understanding that achieving a balance is crucial in all areas of our daily life.

These same studies also discovered that taking a nap contributes positively to our physical and mental health. However, for this to be effective, it should last for at least 30 minutes but be no longer than 90 minutes, as then the nap could harm our sleep that night. Also, a midday nap should not be used as a substitute for night-time sleep, only as a welcome addition to it.

To transition from wakefulness into sleep, we need to be in a state of deep relaxation, which is reflected in a slowing down of activity in the cortex. Difficulty in making this transition – when we find it hard to fall asleep – is the most common form of sleep disorder. Modern medicine and the findings of 21st-century brain research have provided us with tools to deal with this problem. The tool that people reach for most often is the sleeping pill.

Sleeping pills contain chemical substances that inhibit the activity of the cortex and so bring on the calm state necessary for falling asleep. The problem with taking them is that they can be addictive. The more someone gets used to them, the less effective they are, so it is necessary to increase the dose. I prefer to use natural methods to help bring on sleep. However, when these do not work, normal sleep achieved with the help of pills is preferable to sleep disorders, but should be carefully monitored.

For those who want to use some of the natural tools available to bring relief from sleeplessness, the first thing to do is make necessary adjustments to the environment in which they are trying to sleep. It is important that the bedroom is for sleep only, so it needs to be free from items related to activities they are busy with when they are awake. If there is a TV, desk and computer in the bedroom, they need to be removed. The brain algorithm will then be freed to prepare the brain for sleep without distractions. Ensure that there is as little light as possible, perhaps installing blackout blinds.

While total darkness is ideal, if lighting is needed, it is important to avoid blue light. This colour of light has the shortest wavelength and the most energy of the colours in the visible spectrum. When absorbed by the eyes and transmitted into deep areas of the brain, it creates the impression of daylight. So it inhibits the brain's slowing down in readiness for sleep, which is an essential part of falling sleep. Blue light also impedes the secretion of melatonin, which is one of the essential processes involved in effective sleep. The best type of lighting is monochromatic, such as a red, green or yellow bulb – anything but blue.

Scientific support for the importance of darkness has come from the observation that there is a higher prevalence of breast cancer in women who regularly sleep with lighting or a TV on.[12] The researchers were helped to decipher this puzzling phenomenon by gathering insights from brain research, then explaining it as follows. During sleep, the eyelids open from time to time without us being aware of it. When lights are on, light penetrates the eyes and reaches the deep brain, which concludes that the night is over and stops the secretion of melatonin. As a result, there is a decrease in the brain's ability to monitor vital processes in the body, so things are missed or not done, including the elimination of newly formed cancer cells.

Also recommended is avoiding physical activity and food shortly before bedtime, as maximum relaxation needs to be achieved to aid falling asleep. Things to avoid include drinking coffee and consuming other stimulants. Also, watching television and looking at other screens are problematic because of the effects of the blue light emitted from them. Activities that involve excessive use of the cortex, such as thinking, memory and concentration, should also be avoided.

So, in summary, slowing down and limiting your activity as much as possible prior to going to bed should enable you to get off to sleep. For people who need extra help getting into the zone, there are various options, such as listening to recordings of guided relaxation, white noise, waves lapping on a beach, gentle rain and so on.

That just about sums up the medical recommendations and, in principle, they are all valid. Nonetheless, because the brain is flexible, researchers do not unequivocally accept the medical recommendation that we should avoid any loud activities before bed. For example, there are people for whom watching television is calming, allowing them to slip into a pleasant sleep. For these people, a real connection has been formed over time between watching TV and a slowdown in mental activity (a conditioning process has occurred). For them, many brain researchers would recommend that they keep on watching TV before going to bed, but only on the condition that it is turned off (either by a timer or someone else), immediately after falling asleep. For those who need or want to stop watching TV to fall asleep, thanks to the flexible brain, connections can be developed over time between other specific activities and a slowing down of the mind.

Disturbances to sleep may be caused by us suddenly waking up in the night, often because of a need to visit the bathroom. If the awakenings are few in number (no more than three, say) and followed by a quick return to sleep, this is not a serious problem requiring treatment. But if there is difficulty getting back to sleep, it may be a sleep disorder and should be treated using the methods described earlier in this chapter and a visit to a doctor if it persists.

Sleep apnoea

Sleep apnoea is a potentially serious sleep disorder in which a person repeatedly stops and starts breathing. Apnoea means temporarily stopping breathing. This causes multiple brief awakenings during the course of a night, sometimes ten an hour, of which the sleeping person is completely unaware.

As we have learned in this chapter, cells in the body are dependent on a regular supply of oxygen. This comes from air inhaled into the lungs, which is absorbed and travels through blood vessels to cells in the body. The brain supervises this process.

In normal circumstances breathing does not stop, even during sleep. Some people suffer from a contraction of the airways, a phenomenon that tends to worsen when they are in deep sleep (this is also the cause of snoring). When this narrowing of the airways leads to a complete blockage – which may happen during deep sleep – and breathing stops, the area of the brain responsible for managing breathing locates the malfunction and wakes the sleeping person up so they can breathe normally again. Each awakening lasts only a few seconds, without the person being aware of it.

This sequence of events – falling asleep, a narrowing and blocking of the airways, apnoea, rapid awakening, the regulation of breathing and returning to sleep – is repeated throughout the night. As you might expect, people with sleep apnoea feel very tired during the day, even though they have slept for the number of hours that, normally, would result in them feeling rested. But fatigue is only the tip of the iceberg. When it is acute, this sleep disorder damages the functioning of the brain to the point of cognitive decline, and increases the risk of heart attacks and stroke.[13] This is why when doctors hear complaints from patients that they are experiencing constant fatigue "for no reason", this will raise their suspicion that the cause could be sleep apnoea and they will refer them to a specialist for a diagnosis. If the suspicion is confirmed, investigations will be undertaken to locate the cause of the narrowing of the airways. There might be a physical cause, such as a polyp at the back of the nasal cavity. Sometimes the only treatment required might be sleeping on their side, abstaining from alcohol or quitting smoking. If none of these help, the effective way to enable normal sleep is to use a device called a continuous positive airway pressure (CPAP) machine, which delivers air under pressure to a mask to keep the upper airways open.

Some tips for improving your sleep

Being able to fall asleep at the end of the day depends on two components:

1 feeling tired
2 a period of calm relaxation before lying down to sleep

Assuming that you are tired by the time evening comes, let's focus on some simple techniques you can use to create a state of calm relaxation.

1 Direct your thoughts inward and concentrate on a moment or situation in which you experienced calm. Allow the sights and sounds of your pleasant memory to come into your mind and you will soon recreate the pleasant emotion you experienced originally.
2 This time, direct your attention outwards, to your surroundings – watching TV, reading a book, listening to a podcast or music. Any of these can induce a welcome calm that can help you to fall asleep. Take care that the external stimulus (light, TV or audio device) will turn off once you're asleep.
3 Now, combine the first and second strategies by listening to a guided relaxation or meditation to enter a calm, relaxed state. In this method, the external stimulus directs the internal attention to slowing down.

***

If you are in a state of high emotional arousal before bedtime, rather than fight it, simply perform some tasks that will help you to regulate your emotions in a natural way, bringing them back down to a normal level.

Experiment with different techniques to discover the ones that work best for you.

CHAPTER 11
CAUTION: DANGEROUS FOOD

The general sophistication of the human brain – a result of it adapting to the many changes of circumstances the human species has faced during its existence – is not evident when it comes to our eating habits. This part of the brain's activity has not changed much at all. The scarcity of food that haunted our ancient ancestors led to the creation of defaults that were essential for their survival, and these are still with us today. We see them at work when we eat as much as we can or crave sweet and fatty foods.[1] The failure of the human brain to adapt to the new reality of food being generally easily available in the modern world has caused some of the distortions that we see in our current eating habits, both in terms of what we eat and the quantities consumed.

Why we eat and what we need

Before we go into this subject in more depth, it is important to remember one general point: eating and drinking are not meant to give us pleasure. Their one and only purpose is to replenish the stocks of chemical substances in the body that make up our cells, which are being repaired or replaced all the time. The raw materials our cells need are proteins, sugars, fats, vitamins, water and various minerals. The energy produced by digesting the food to retrieve these materials also enables the removal of waste.

Knowing this, we then need to ask ourselves these two questions. How much food is necessary? What foods should we eat and in what proportions?

Are you hungry?

Regarding the first question, as to how much food is necessary, when I studied medicine in the 1990s, the prevailing belief was that an adult needs, on average, 2,500 calories per day. The exact amount would be affected by variables such as gender, body weight and daily physical activity. However, we have come to realize that this is too much – a conclusion drawn from observing eating habits in countries in the Global South, where people enjoy robust health and good life expectancy. Researchers noticed that people in these parts of the world consume smaller amounts of food than those in the West. It seems that in the Global North, and in the West in particular, people too readily accept the mantra that we should be having three square meals a day – morning, noon and evening. In fact, one meal a day – or even one every two days – is all adults need to feel satisfied, if it contains the correct amount of food the body needs for that period of time. What has come to be perceived as a *need* for three daily meals, often supplemented with snacks stems from a false feeling of hunger generated by the relevant area of the cortex. False because the body does not need the additional raw materials. Hold that thought for now – there will be more on the ramifications of this messaging in the brain later in this chapter.

Fat: good or bad?

For the second question – about what foods we should eat and in what proportions – we need to go back several decades to the Seven Countries Study. Its conclusions caused far-reaching changes to eating habits in North America and Europe.[2] Scientists led by nutritionist Ancel Keys looked at the diets of seven countries with different cultures, and tested two variables: the amount of fat residents ingested daily and their life expectancy. The findings were unequivocal. In countries where the staple foods were rich in fat, especially saturated fat, the incidence of heart disease was higher and life expectancy shorter. In the blink of an eye, low-fat diets came into being and gained traction around the world, but that was not the whole story.

Years later, scientists revisited the Seven Countries Study, as various myths had arisen about how the results were arrived at and what the true meanings of the results were.[3] Essentially, it was found that Keys was not suggesting a low-fat diet should be followed but, instead, recommended a Mediterranean dietary pattern as being the best for optimal health in the long term – that is, one in which levels of saturated fat and animal products consumed are fewer, and there is a higher consumption of vegetables, fruits, whole grains and nuts than is still usual in the West.

There has been a change in the way we perceive the value of fat in our diet. It is widely accepted that fat is very important to the body and low-fat diets can cause severe damage.[4] Furthermore, no connection has been found between eating saturated fat and heart attacks.[5] Yet, we are all aware of how the aversion to fat has found its way into recommendations made in the media that leave us with the impression that *all* fat is bad for our health. Instead, obesity and other chronic illnesses have multiple causes. It is the consumption of excess calories, high intakes of animal products, refined sugar, not being physically active, smoking and other lifestyle factors that have the most impact on our health.

Sugar is not our friend

The fear of fat diverted attention away from the real enemy that threatens our health: sugar.[6] There is no doubt in terms of food that sugar is the number one cause of heart attacks, diabetes and obesity. Unfortunately, the human race is addicted to sugar.[7] One of the conclusive pieces of evidence for this is that eating sugar and snorting cocaine activate the same areas in the brain.[8]

The main danger lies in sucrose – the white or brown sugar refined from sugar cane and sugar beet that finds its way into most processed foods, sweets or candies, desserts, pastries and soft drinks. Sweetness is also found in fruits and honey, but eating them in a balanced way will not harm the body. On the contrary, they are a good source of essential vitamins and minerals.

The sugar component that *is* needed by the body can be found in foods that are not necessarily sweet, such as grains, legumes and potatoes. Unlike sucrose, which is a simple carbohydrate, the sugar in all these foods is a complex carbohydrate.

This next question will surely cause you to raise an eyebrow. Which is more fattening, one calorie of fat or one calorie of sugar? The answer seems crystal clear, so you probably said, "Both types of food are equally fattening as they have the same number of calories."

For many years, this was the prevailing opinion among scientists as well. Only more recently has it become clear that things do not work like that. Fat and sugar do, indeed, have the same number of calories, but the way the body processes them is different. A large proportion of calories from fat are converted into energy, so they dissipate from the body. The fate of calories from sugar is quite different. On entering the body, the pancreas secretes the hormone insulin and converts most of the sugar into fat, which is stored in the fat cells. The end result is the answer to our question: sweet foods are more fattening than fatty foods.

Tasty!

Taste plays a big role in our eating habits. The sensation of taste begins in the mouth – the place where the body first comes into contact with a food. Sensors scattered around the tongue, palate and inside the cheeks detect flavours and transmit them through nerve endings for processing in the areas of the brain relating to taste. The purpose of the processing is not so that we may enjoy our food but to enable us to survive in the world. This is because our brain, even now, is still programmed to be in hunter-gatherer mode. It only needs a limited number of basic tastes to ensure that we consume edible food and shy away from inappropriate, even dangerous, food.

Sweetness is an indicator of a food's ability to provide immediate energy. It was essential to our forebears when they were chasing food on the hoof or running away from predators. The brain was not content with only creating sensors to detect it but it also forged an internal connection between sweetness and pleasure to create the urge in us to consume sweet foods.

Saltiness in food indicates the presence of minerals such as magnesium, calcium, zinc, iron and sodium, all of which are essential to the body. But they need to be in the right doses. For example, excess sodium causes an increase in blood pressure, so the brain has created a reluctance to eat excessively salty food as a warning. The problem is that our bodies tend to get used to salt, so we end up consuming more of it than is necessary or healthy.

The detection of two more tastes – sourness and bitterness – was also intended to keep us alive but works in the opposite way to sweet and salty tastes. Instead of attracting us, they make us dislike the food, which works to prevent us from eating foods that are not good for us. For example, when we take a bite from an excessively sour fruit, the taste makes it clear that the fruit is not ripe or even poisonous, and if meat has a bitter taste, this indicates it is going off and contains harmful bacteria.

At the beginning of the 20th century, Japanese researchers identified another flavour and gave it the name umami, which is Japanese for "delicious". This relatively new flavour has no precise definition but is a kind of savoury, meaty taste. The umami receptors allow the brain to recognize proteins that are crucial to the body and its functions, and these are coded in the brain as nutrients, although they are of less importance than sugar and fat. As they are not vital, unlike sugar and fat, umami tastes do not activate the pleasure areas of the brain. They are, though, very tasty to us when balanced with other flavours.

In Japan in 1908, biochemist Kikunae Ikeda concocted a certain powder.[9] He was trying to isolate and duplicate the taste of kombu – an edible seaweed used to flavour a popular Japanese broth called dashi. It contains the protein that activates the umami taste sensors. That powder was monosodium glutamate, and it has been used ever since as a controversial flavour enhancer in food manufacturing and various fast foods, including takeaway Chinese food.

At the beginning of the 21st century, another taste was identified that developed during our evolution: the taste for fat.[10] In the time of our ancient ancestors, it served the important purpose of encouraging them to consume fats as a store from which energy could be created to keep them alive.

You might be wondering (as in chapter 6) why I left spiciness out of this list of flavours – after all, it is the taste that makes so many dishes worthwhile. While I certainly enjoy hot, spicy food, I did not include it as it is not categorized as a taste. This is because it is not detected in the same way as the other tastes we have been looking at in this chapter. The sensation of spiciness is created thanks to the activation of pain sensors in the mouth rather than taste sensors. If you want to prove this for yourself, rub a raw chilli on your hand and you will feel a burning sensation. The reason for this is that it contains a substance called capsaicin, the active ingredient in all types of pepper, which activates pain sensors found throughout our body.

When we eat, the relevant areas of the brain perceive not only the basic taste of the food but also its smell, texture and appearance. All these aspects are processed in the cortex and the output regarding the combination provides us with the flavour and other information about our food. This ability to taste originally developed to enable us to determine the quality of food in terms of its contribution to survival but has since evolved substantially.

From running on empty to feeling full

Feelings of hunger and satisfaction regulate the amount of food we eat. Once food has passed through the digestive system, the stomach and some other organs secrete a hormone called ghrelin. When ghrelin reaches the brain, it activates the hunger centre in the hypothalamus. As you will recall from chapter 3, the hypothalamus is a structure located in the subcortex that is responsible for, among other things, the behavioural and physiological expression of emotions.

The role of the hunger centre is to make food relevant to us. It does this by creating a connection between food and the experience of pleasure. Thus, every thought about food stimulates the pleasure zone in the brain and, as a result, the desire to eat increases. The opposite happens when a sufficient amount of food accumulates in the digestive system. Then, a hormone called leptin is released from the fat cells in the body.[11] It also reaches the hypothalamus, but activates a different neural network called the satiety centre. This interrupts the cycle linking food with pleasure and causes a feeling of being replete.

However, this description of the regulation of our hunger and satiety centres does not stop there. The flexibility of the brain has allowed it to create various ways to bypass these mechanisms, such as the false feeling of hunger mentioned at the beginning of this chapter. It did this to ensure that our ancient ancestors survived as food could often be scarce. These mechanisms continue to operate even in today's world where, for most of us, such periods of scarcity do not happen.

A quick visit in our imagination to a group of our hunter-gatherer ancestors will clarify this process. On their wanderings, they come across a tree loaded with ripe fruit. They sit by it and enjoy the fruit. After a while, their stomachs fill up and the leptin hormone begins to work at full strength. At this point, we would expect them to stop gorging, but they keep eating. This is because their nutritional insecurity – a fear that if they return to the tree tomorrow, they will find that the remaining fruit has gone – led, over time, to the creation of a way to bypass the effects of leptin.

In terms of promoting the survival of the human race, this evolutionary development served early humans such as our hunter-gatherers well. They did not know when such treasure would come their way again, so from a survival point of view it was right for them to carry on eating to accumulate a large amount of sugar in their bodies to be stored as fat and used later in times of need.

The existence of this way to bypass the satiety centre is known to most of us through experience. Imagine you are enjoying a hearty family dinner. After eating your fill, your host comes out of the kitchen carrying a chocolate cake topped with whipped cream. Even though your satiety centre is giving you a clear message that you are full, after being triggered by the release of leptin earlier in the meal, you will likely devour your slice of the cake, down to the last crumb.

Research has shown that satiation bypasses evolved for sugar and fat. There are two types of food that could regularly be in short supply for our ancient ancestors, yet are easy for the body to convert into stores of energy.[12] These bypasses have remained intact despite the transition to an affluent lifestyle all these many years later. So we devour the chocolate cake even though we are already full and have access to more cake. The bypassing of the satiety centre in this case was caused by information sent by the eyes to the brain when the cake arrived at the table, but our other senses can also be the source of this effect on us.

Our emotions can also play a part. Raiding the freezer for ice cream or eating chocolate are familiar responses to feeling sad or stressed, as eating them generally improves our mood. This is because food can create an internal connection to pleasure, especially when the food is sweet. We saw this in the discussion of different tastes earlier in this chapter because of its potential to contribute to our survival. This phenomenon is known as emotional eating, and it happens because the natural bypasses prevail over leptin's hormonal cry of, "Enough, you are full!"

Reprogramming the brain

As we have seen, the feelings triggered by the hunger and satiety centres that influence our behaviour were originally set up to meet the nutritional needs of our ancient ancestors and do not reflect our needs now. Fortunately, as the brain is flexible, it is able to overcome these ancient impulses and emotions. It can settle instead on a lower intake of food that still supplies the raw materials needed to repair ongoing wear and tear in our brain and body. We have the mental tools to adjust our intake, deciding which foods to include and the amounts we consume to meet our actual, rather than historic, physiological needs. Of course, such changes can't happen overnight. It's a learning process, requiring motivation to succeed, but it is definitely possible. Deepening our learning about food reprograms the brain, switching from its default patterns that served hunter-gatherers to a mode of impulses that are more suited to modern life.

Eating disorders

Eating too much or too little can cause serious harm to our health. Even when such patterns arise from deficiencies in the regulation of the hunger and satiety centres or environmental influences, they are not considered eating disorders, no matter the damage caused. It is only when they stem from mental health problems, such as poor self-image, depression or anxiety, and are not related to appetite, that clinicians call them eating disorders. The two most well-known disorders are bulimia and anorexia, and they are the most dangerous and difficult to treat.

Bulimia

Bulimia is an eating disorder involving bouts of binge eating that are motivated by the search for pleasure. During such a bout, a person is able to eat everything they can get their hands on, consuming up to about 10,000 calories. One young woman told me that during a bulimic attack she could, without difficulty, consume the entire contents of her fridge and then start on the frozen items. Once, while trying to swallow frozen chicken, one of her teeth broke.

Unfortunately, the damage that can be caused by bulimia can't always be fixed by visiting the dentist. Eating such huge quantities of food can lead to intestinal and stomach ruptures. The purging of the food that follows these binges, by vomiting, laxatives and other methods, causes dehydration and mineral deficiencies, and frequent bouts of bulimia can even be fatal.

As noted, eating disorders can be associated with a poor self-image. The pattern for our self-image in the brain is a neural network with two arms – one arm in the cortex and the other penetrating through the cortex and into the subcortex area. Low self-esteem leads to various everyday situations being perceived as threats, expressed as a heightened tendency to be offended, angry and resentful. People who feel this way suffer more than average. In contrast, people with high levels of self-esteem have a correctly regulated amygdala – the part of the brain responsible for creating threatening feelings – so they have greater immunity to the everyday sensations of threat and, therefore, a greater sense of mental wellbeing.

In some cases, people with low self-esteem discover a workaround to reduce the threats they perceive: they eat. Eating activates the neural network that creates pleasure. The problem with bypassing the body's system like this is that the positive effect is temporary and, after a binge of emotional eating, it falls back into its usual balance. So, once again, the threatening sensations dominate. Sometimes the repetition of this cycle only increases the anguish, pushing the amygdala to work even harder. People with bulimia can sink into a danger zone of compulsive eating, beset by intense and uncontrollable feelings of disgust and self-hatred. Afterwards, all they want to do in this desperate situation is to empty their stomachs of the disgusting food that caused these feelings as quickly as possible.

The proportion of women with bulimia is nine times greater than that of men, and the phenomenon is not rare. A comprehensive survey conducted in the USA found that approximately one per cent of all young women have attacks of bulimia, and noted that the frequency of the attacks is not constant.[13] Symptoms of bulimia can also be found in people who do not binge eat, then induce vomiting or take laxatives.

Historically, it was common to treat people with bulimia with antidepressants and anti-anxiety medicines but, in most cases, there was little proof that these were effective. A significant breakthrough in treatment has been achieved with the help of cognitive behavioural therapy (CBT), with recovery rates reaching levels as high as 50 per cent after 12 months of treatment. CBT can be effective as it is possible to address the fact that the urge to eat in bulimia comes from a longing for pleasure, not hunger. An understanding of the cognitive aspects of bulimia allows the brain to create activity in the cortex that has the power to correct this distortion. The success of the treatment depends on the extent of someone's motivation to be free from bulimia.

Anorexia

At first glance, anorexia is the opposite of bulimia – a morbid aversion to the intake of all food. It is certainly more severe than bulimia, as it is the result of a distortion associated with a person's self-awareness. Similar to bulimia, anorexia is an eating disorder that mainly affects women. Two-thirds of those with anorexia are young women, most of whom attach supreme importance to thinness. Often, they see the frail, fashion model or Barbie doll-type figure as an ideal example of beauty.

Most of us, even if we think a certain idea is very important, do not regard it as something worth living for, and especially not dying for. The ability to develop such a vibrant interest in an idea requires a fairly high level of intelligence. Indeed, many of those with anorexia are intelligent young women, who have fiercely strong willpower that pushes them to reach any goal they set for themselves, no matter what the cost.

Anorexia develops in stages. It begins with a gradual reduction in the amount of food eaten. At this point, the person is elated at their success in doing something that they see as very significant. The results soon show themselves – the person starts to lose weight. As the process continues, their sense of satisfaction increases. At the same time, the meaning they attribute to thinness takes over all the functions in the cortex and subcortex. The brain has internalized this message and is now entirely mobilized to fulfil a flawed and erroneous goal. When normally there would be a feeling of hunger, this gives way to an aversion to food. The fear of getting fat does not leave this person alone. They have stopped eating and are frantically looking for ways to hide this from the people around them. To speed up weight loss, they engage in excessive physical activity, sometimes in a way that is so out of control, they continue to the point of complete exhaustion.

At this stage, the person with anorexia does not feel in any way satisfied. On the contrary, the annoyances and anxieties they associate with their weight only intensify as they become thinner. When a mirror is placed in front of someone with anorexia who can't safely get any thinner, they will still see themselves as being too fat. This distortion of their perception of reality was created when the idea of wanting to resemble a model or Barbie doll took root in their cortex and, as a result, became an overarching target that their entire brain was mobilized to achieve.

According to the clinical definition of anorexia, it is a psychotic condition that requires enforced hospitalization as there can be complications resulting from this disorder. Hospitals have specialist wards for those with anorexia, the treatment including cognitive and other therapies and naso-gastric feeding. Even so, recovery can be a very difficult process and there is no guarantee of success because those with severe anorexia value thinness over everything else. I recall seeing a weak young woman in hospital, holding a piece of chicken on a fork and, despite the encouragement of the caregivers, she was unable to put it in her mouth. A doctor described how hard trying to eat is when you have anorexia and I imagine it sums up the feelings that young woman must have been experiencing in that moment: "Imagine a Jew being ordered to swallow some pork on the Sabbath."

The importance the brain attaches to thinness is a barrier to the possibility of a medicinal "miracle cure" for anorexia. The only effective way to help those with anorexia is CBT, as it is for bulimia – the success rate of such treatments reaching levels of about 30 per cent.[14] This is a remarkable achievement considering the severity of the disorder. The goal of CBT for anorexia is to help the person to create a new purpose in the hope that it will overcome the purpose that initiated anorexia.

To illustrate how this can work, I will now describe the way in which it helped Rebecca.

Rebecca was a postgraduate science student, trapped by severe anorexia that was seriously disrupting her life. Her supervisor at the university, who initially had high hopes for Rebecca, lost all interest in her. Rebecca also gave up on herself.

After being hospitalized several times at the request of Rebecca's parents, and always slipping back into her old ways, it was decided as a last resort to try CBT. During treatment, it became known to Rebecca's psychologist that, in her youth, she engaged in a lot of physical activity, especially in the gym, and really enjoyed it. Armed with this piece of information, the psychologist talked to Rebecca about the wonders of fitness training and how it's great for toning the body and strengthening muscles.

The idea slowly entered Rebecca's awareness and motivated her to take up fitness training again. In the process, she also realized that if she didn't start eating, she would lack the nutrients necessary for physical exercise. Thus, she found herself standing at a crossroads where she had to decide between eating to attain the body she wanted or continuing to avoid food for the purpose of promoting the same goal.

After a prolonged period of treatment, the new motivation prevailed over the one that had caused her anorexia. Rebecca successfully completed her postgraduate studies, and now you can find her working out three times a week in the gym in the healthy way that many people do. Most importantly, she eats sensibly without giving it too much thought. Rebecca talks with complete candour about her anorexic period as "past madness".

Unfortunately, such success stories are rare. Around half of those undergoing CBT will still think non-stop about food, forever count calories and weigh themselves incessantly. These unfortunate individuals will continually be underweight and have to deal with the condition in all its severity. Yet, as Rebecca's story illustrates, other positive outcomes are also possible thanks to the flexible brain and the ways we can harness it to help us with our mental health.

Some tips for avoiding dangerous food

There is an overwhelming consensus among researchers that refined sugar is associated with health problems, and this holds true for both white and brown sugar. Here are two approaches you may like to try for a sugar detox that are based on science.

Detox one

1　Decide that, during the next two days, you will not eat foods containing sugar. On each day, focus on the availability of foods that do *not* contain refined sugar, but allow yourself to eat natural honey and fruits.
2　At the end of the second day, emphasize the sense of achievement to yourself. Your brain will reward you with a wonderful feeling of pride and self-esteem, and boost the level of your motivation to continue detoxing for two more days.
3　At the end of the fourth day, once again recognize your achievement and take pride in it. Initially, your brain will be motivated to respond to the challenge of not eating sugar because of the feelings of pleasure and achievement experienced at the end of the second and fourth days of abstinence. Later, the brain will register this feeling as an existing motivation and the learning that has been created as a result of the notion that foods containing refined sugar are no longer necessary because you have replacements you enjoy.

Detox two

1　Here is another way to reduce consumption of refined sugar. This method is called aversion therapy, which is a way to put yourself off something that is not good for you. Everybody can change the way the brain automatically links pleasure to eating sugar by getting it to link sugar with a threat instead. A central condition for this method to be effective is establishing real motivation from within to avoid eating sugar.
2　Let's start. Imagine the following scenario and concentrate on it. On a sugar plantation in a poor area, workers are harvesting sugar cane. Focus on one of the workers. He is feeling ill and he vomits all over the canes that have already been cut and are on their way to make sugar. Revulsion, if it is felt now, will disappear in a few seconds but repeatedly visualizing this scene, accompanying it with the feeling of revulsion, should help it to take hold and later make foods containing

sugar abhorrent, or at least less attractive to us because we now associate it with unpleasantness, not pleasure.

***

We know now how we can use the brain code to leverage unpleasant sensations for worthy purposes. In chapter 12, we will see that sometimes, following a disruption in our natural mechanisms that regulate our emotions, unpleasant feelings caused by fear and anxiety may overwhelm us. We will learn what the characteristics of a feeling of threat are and how to re-engage those natural mechanisms, harness them and take back control of our mood, so we can think more positively once again.

CHAPTER 12

FEARS AND ANXIETIES

We tend to talk about fears and anxieties in the same breath, for good reason. Both these emotions cause the amygdala to create an overall sense of being threatened. Yet, despite the amygdala being the common denominator, the way we experience fears is essentially different from the way anxieties make us feel. Fear is the result of a stimulus reaching the subcortex. Usually, that stimulus comes from the outside world, but it can also come from within us, such as physical pain. Anxiety is not the result of a stimulus from something physical, but a type of response to a threatening scenario that is created in the cortex. If a vicious dog is heading toward me, I will be gripped by fear, which will make me run away. However, if I'm about to wander into an unfamiliar forest after dark, it is likely that a feeling of anxiety will come over me and affect how I proceed. In both these situations, the amygdala was activated as part of the brain's concern for my survival, but the outputs were different.

Fear and anxiety are natural emotions that are an essential part of our response to situations in daily life. But, for them to fulfil their role and be of help, one important condition must be met. The response of the amygdala should be appropriate to the reality of the situation. The problem is that this is not always the case. Sometimes the amygdala's response sets off an alarm that is at an inappropriately high level, so it reverberates through the brain when there is no real reason for such a reaction. The opposite can also happen – the amygdala's response being one of indifference to a threat when it should have provided more of an alert. This is not unusual as the amygdala works efficiently in only a small minority of people who have highly developed emotional intelligence. In this chapter, we will focus on understanding what is happening when there is permanent imbalance in the functioning of the amygdala. This is a condition that characterizes people with psychopathic personalities.

No fear

Psychopaths tend to feel less threatened than the average person, and this lack of fear means that they are not averse to taking risks. Many criminals have psychopathic personalities. At the time of committing the crime, they were aware of the possibility of getting caught, but didn't feel threatened by this prospect, so it did not deter them.

However, the underactivity of the amygdala that is behind such a response is not always a drawback. Studies have found that people whose amygdala is less sensitive than usual may excel at highly stressful jobs, so may make excellent surgeons or fearless firefighters.[1] Although such people are aware of the dangers and magnitude of the responsibility assigned to them, they are able to act more calmly than the average person thanks to their dormant amygdala.

Fearlessness also characterizes what researchers call the "adolescent brain". This covers young people in their teens and early 20s. In this age group, the frontal cortex – the area that relates to cognition and planning appropriate, rational responses, including anxiety – has not reached full maturity. So it does not activate the amygdala in the same way that would generally be expected in a balanced adult. The behavioural expression of this is teenagers taking risks and getting carried away without thinking twice about where their adventures might lead. For example, a young motorcyclist knows that speeding down a winding road might end in a crash, but this will not make them anxious, so will not put them off doing it. On the contrary, the excitement that they feel at the prospect will spur them to go faster, without a thought for the potentially disastrous outcome. If we now fast forward to quite a few years later, and the not-so-young motorcyclist happens to recall this mad dash, they will probably look back in horror that they did such a thing! The difference in thinking is because, once the frontal cortex has fully matured, it allows the person to understand the seriousness of what they did, so the thought of it now prompts a sense of threat from the amygdala.

Neuroscientists have come to understand that this adolescent brain stage of development was a necessity as humans evolved.[2] Millions of years ago, it was the fearless youngsters, thanks to their half-developed brains, who persuaded their worried parents to abandon depleted hunting grounds and look for richer sources of food. The migration of early humans from Africa to Asia and Europe was also made possible to a large extent by the adolescent brain, as young people would not have been afraid to set off on an adventurous journey into the unknown.

The adolescent brain can spur young people on to considerable achievements, but might, at times, result in destruction and mayhem. It also gives them the ability to implement bold ideas and break new ground that older adults, who may be beset by fears and anxieties, shy away from. Equally, it can give young people a potentially unrestrained destructive power. Education, which imparts ideals, plays a central role in directing the adolescent brain down one path or the other.

Fear and anxiety

Panic attacks

Underactivity of the amygdala results in the absence of fear and anxiety. Overactivity results in the opposite situation, with high levels of fear and anxiety felt for varying periods of time for no good reason. Such fears and anxieties can become intolerable. The most extreme state induced by such overactivity is a panic attack. One patient I spoke with who had experienced panic attacks described what they were like:

> These were the longest moments of my life. They went on forever. I felt a terrible horror. I could see clearly that the world around me had been destroyed and I was going to die.

This is only a partial description of what happens during such an attack. The amygdala is not satisfied with simply sounding the alarm, bringing a feeling of anxiety into a person's awareness. It also activates large areas of the brain, resulting immediately in perspiration, shortness of breath, a very rapid pulse, dizziness and incessant movements of the muscles. All this consumes a lot of energy so to prevent exhaustion that can impair survival, the brain does not allow an attack to last for more than a few minutes.

I have experienced one panic attack in my life. It happened a few hours after I learned that I'd got a place at medical school. You're probably asking, "How come?" Well, it also struck me as strange to begin with, but when I thought about it much later, I understood what had happened.

The news was indeed good, but it also caused me to feel concerned that maybe I'd find the coursework too hard and what my friends and family would say if I dropped out. It felt like a rush of lots of overwhelming thoughts. The higher-order regions of the brain, which processed my fears, alerted the amygdala and its response was to judge them, unjustifiably, as indicators of a complete disaster. In simple terms, the amygdala created a mountain out of a molehill.

In my story, there was a reason for the panic attack, but this is not always the case. An attack can strike someone at any time, anywhere – while they're driving, watching TV, enjoying some sun by the pool or any number of other situations in which it would be completely unexpected. Everything seems fine then, suddenly, for no reason, the amygdala interprets a message as a threat and decides to respond forcefully, with no justification. Should such attacks become frequent (some people experience them several times a week), this would fulfil the criteria for a diagnosis of panic disorder. People with this disorder may develop agoraphobia (a fear of open or crowded places) and so shut themselves in their homes.

Post-traumatic stress disorder

Another severe condition in which overactivity of the amygdala features is post-traumatic stress disorder (PTSD). People develop this condition following a situation in which there was a threat of some kind, ranging in severity from events that would not concern most people to very serious ones in which they could have died, such as a car accident, rape, armed robbery or war. Most people who have gone through such difficult experiences manage to return to normal life, but there are some, including war veterans, whose traumas continually recur as tangible flashes and cause them nightmares. They are afraid to do anything remotely connected to whatever horror they have gone through and are in a state of constant terror.

This miserable condition is caused by the amygdala working too vigorously during the original experience. As a result, the event registers as a traumatic memory in the amygdala, and it does not stop replaying it. From the brain's point of view, the process of memory accompanied by increased activity of the amygdala is important to our survival, so the amygdala gets the green light to recall memories – good and bad – again and again.

Phobias

The amygdala is chronically more active than it should be when someone has a phobia. A phobia is a constant, disproportionate and usually nonsensical sensitivity to a certain object or situation, expressed as a feeling ranging from fear to abject terror. According to some studies, around 20 per cent of the population is affected by phobias to varying degrees.[3]

We can develop phobias to anything we encounter in our daily lives: reptiles, heights, thunder and lightning, dirt, dogs, cats, darkness, pedestrian crossings, enclosed spaces, open spaces, driving, dentists, needles . . . the list is endless. As mentioned, the feeling is disproportionate to the threat in such instances. For example, we all know that there is a possibility of mechanical failure when we board a plane. But most of us do not let the thought of it prevent us from flying and are calmed by the understanding that the probability is extremely low. But a person with a phobia of flying will feel uncontrollably threatened by the thought of an accident. They will therefore not only avoid air travel, but the very sight of an aircraft, even when it is on the ground, will trigger feelings of fear.

Researchers in the field of evolutionary biology discovered that this phenomenon has a genetic component originating millions of years ago when our ancient ancestors lived surrounded by dangers such as predatory animals and venomous reptiles and felt threatened by the forces of nature.[4] The fears they developed in this wild environment were valuable as they helped them to escape bites, stings, infections, poisonings, falls from high cliffs, drowning in flowing rivers and similar disasters. These fears were therefore preserved and remain within us to this day. For a significant proportion of the population, they have intensified over time. It is interesting to note that fear of snakes, spiders, heights, thunderstorms and open spaces or confined places are still among the most common phobias.

Sometimes a phobia can be generated by an event that ended badly. For example, a person who has been bitten by a dog may succumb to fear whenever they see a dog because of the memory of that original experience, even if it is a puppy on a lead. It is also possible that witnessing an event, rather than being involved in it, will result in a phobia developing. For example, seeing a pedestrian run over by a car may lead to a fear of crossing roads, even when there are traffic lights.

Anxiety

Less severe than most phobias but more commonly experienced is general anxiety. This is an acquired condition that develops over time and is indicative of overactivity of the amygdala. The cause of general anxiety is not a congenital defect in the amygdala but a sudden change, usually following a specific event, that makes it more sensitive than it was previously.

The intensity of the activity in the amygdala during a period of general anxiety is not as great as it is during a panic attack, so the brain allows the situation to carry on for longer. Yet because an overactive amygdala intensifies the feeling of being threatened, this condition of general anxiety may also lead to behaviour that is more stubborn and aggressive than the circumstances require. Among other things, this will manifest in a tendency to jump straight to a fight response when a better course of action would have been to freeze or, perhaps, graciously withdraw. Essentially, as people with general anxiety see reality in a gloomier and more threatening light than people blessed with an amygdala that is functioning normally, this dictates their behaviour.

Some tips for regulating anxiety

As we saw in chapter 8, mirror neurons enable us to synchronize our activity with those around us. For example, someone acting in a calm, peaceful and relaxed way can have the effect of calming other people nearby, so they will start to feel and act in a similar way. However, this ability to influence behaviour is a double-edged sword because frantic and anxious people can bring about the negative emotions of anxiety and frenzy in others around them. So if you are beginning to feel anxious, it is best to avoid being near someone behaving in this way.

How does this transfer of emotions take place? The external mediators of the communication of emotions are our facial expressions, body language and intonations of the voice.

Here are a couple of ways to put what we have learned into practice when you experience anxiety.

Method One

1 Take a look around you. If you are experiencing a negative and unbalanced emotion of frustration, anger or jealousy, it will be beneficial for you to be physically or mentally close to a calm and peaceful person – someone who is in a positive, balanced emotional place.
2 Identify a calm person by reading their body language. Once you have, you do not need to talk to them; just being close to them will do the job.
3 Let your mirror neurons and their calming effect work on the negative emotions you are experiencing and they should be pleasantly reduced.

Method Two

1 Take or download a photo of someone who, in your mind, is a good, peaceful, cool-headed and pleasant person and who is adopting a calm expression.
2 Enlarge the image enough to be able to distinguish the facial features and save it.
3 For optimal regulation of your emotions, any time you experience an increase in an unbalanced, negative emotion during the day, call up and concentrate on the photo. Focus especially on the person's facial expression. Notice how the intensity of the negative emotion decreases.

***

Use these and other exercises so you can become more proactive in regulating your emotions (also revisit the exercises at the end of chapter 4).

CHAPTER 13

THE HIGHEST LEVEL OF THE BRAIN'S FUNCTIONING

A good way to understand brain activity is to compare it to a sophisticated industrial complex, with production spread across different sites and floors. Each of the factories is entrusted with the production of a unique component but works closely with all the others toward a common purpose to improve the survival of this whole brain integration enterprise.

Like any enterprise, it needs an infrastructure to enable the machines and workers to work together. The machines are the nerve cells (the neurons), which are built using different chemical substances. The workers are the neural networks, each of which implements a unique brain function, including memory, emotions, concentration and movement (see chapters 2, 3 and 10 for details of these and others). The infrastructure is the brain's ability to provide the nerve cells with nourishment and protection and carry away accumulated waste materials.

For a long time, brain research adopted this analogy of our brain working like a factory, and it is still widely accepted. Only in recent decades have we recognized that the successful functioning of the brain requires another element: an effective manager. As with the industrial complex, having the most up-to-date machines, an efficient infrastructure and a dedicated workforce taking good care of the everyday running of the place will not guarantee a good product if the managerial team is lacking.

Managing the brain: the executive function

Effective management in the context of the brain means applying the higher-order functions in the upper floors of the brain. For example, long-term thinking about how to operate the hardware (including the work environment, the machines and the infrastructure required to operate them) and the efficiency of the workforce. It also means determining and setting the correct priorities and organizing the workers accordingly.

Our "brain manager", called the executive function, resides in a special area located in the prefrontal cortex. This area has been one of the hottest subjects of brain research, and it's easy to see why. Researchers have come to understand that even people equipped with good hardware and diligent workers can find themselves in a problematic situation when it comes to conducting their daily lives because the management is poor.[1]

Prefrontal
cortex

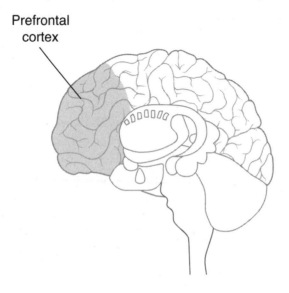

A good illustration of this is people with high-functioning autism (formerly called Asperger's syndrome). As far as memory is concerned, their brain hardware is completely normal and so are the workers. They remember things well but have a serious problem when it comes to memory management. They may be able to remember all the results from the last football World Cup, down to who scored each goal in which minute, but they will have difficulty retrieving a memory relevant to an event in the here and now, such as where they put down their bag just minutes before. There would be nothing to prevent such a person from doing a degree in engineering, but they would probably be stumped by problems that would come up in the daily work of an engineer.

The executive function must simultaneously control the activities in all the various factories that have been integrated into the huge industrial complex. Not only that, but the executive function also needs to be accurate and flexible at all times. For example, when I give a lecture,

I want my working memory (the important information from my memory that I can retrieve into my awareness for a short time) to be related both to the subject of my lecture and relevant examples from my experience at medical school, at the hospital where I worked, at the Weizmann Institute where I did research and so on. But if I suddenly hear a shout, "Doctor, someone has fainted!", I will immediately interrupt the lecture, remember that I'm a doctor and go to help. At that moment, my executive function will be tasked with temporarily storing the memories related to the lecture and, instead, making available to me from my memory pool information about the proper care to give someone who's fainted, which I learned long ago in medical school.

This same executive function oversees various lower-order brain functions, including motor, sensory or cognitive functions or those that control the body. In the motor domain, it is responsible for planning all our movements, according to our immediate needs and those in the future. For example, when we intend to pick up a glass of water, my executive function estimates its size and weight, its temperature and the intensity of pressure that my fingers will need to apply to perform the action. Based on this set of data, it will determine which muscles will be activated and at what level of intensity. Scientists in the 20th century were surprised to find that what were then highly sophisticated robots were unable to lift every cup or plate.[2] The reason for this is that the robot manufacturers at that time did not take into account the importance of movement planning and movement memory. More recently, inspired by brain research, a robotics industry has developed that produces machines capable of learning and remembering moves in a way that is similar to how it's done in the human brain.

A video showing a rotating Charlie Chaplin mask that was discussed in chapter 7 is an excellent example of the role the executive function plays in supervising the senses. It prevents us from being aware of sights that may confuse us. The optical illusion it presents us with is evidence of this at work, as we do not normally see a hollow inside of a person's face when we see them from behind. This help with avoiding confusion developed because, otherwise, it might hinder our survival.

Brain researchers have turned a spotlight onto this area of cognitive management because it is still of vital importance to us.[3] At every moment, we are exposed to countless stimuli. To manage all this brain activity properly, there needs to be a clear distinction between which of these stimuli are important and which are secondary. The executive function, responsible for our cognition, performs this role, deciding whether a specific stimulus is related to our ability to survive or not.

Several other managerial functions derive from this ability to differentiate between items of primary and secondary importance. Among the most important of these are:

- **focusing our attention** – directing the spotlight of awareness onto the important stimuli
- **working memory** – remembering the important things
- **decision making** – the ability to choose
- **emotional intelligence** – the precise regulation of emotions of pleasure and threat

Human beings differ from one another in terms of their ability to distinguish between what is important and what can be ignored. At one end of the spectrum are people with extremely high levels of discernment, verging on genius. In the centre are those with average abilities, while at the other end are people who suffer from a pathological failure in this area, often caused by stroke or an injury affecting the area of the brain relating to the executive function.

To understand how such differences can affect daily life, here's a story a neurologist told me about what happened to a man with a severe pathological failure of his executive function that shows how this affected his ability to sort the wheat from the chaff. The man suffered a stroke and, as a result, the area of his brain relating to his executive function was damaged but otherwise the workings of his brain were unaffected.

One evening, the man was watching TV in the living room while his wife was in the kitchen, chopping vegetables for a salad. At one point, she cut her hand and, in a panic, yelled, "I've cut myself! Everything is covered in blood, even the knife!"

Hearing his wife scream, the man rushed to the kitchen, picked up the knife and washed it thoroughly. Only then did he turn to his wife to wash and bandage her injured hand.

What happened here? The man realized that his wife was injured. He also knew that her bleeding hand needed to be treated, but the pathological failure affected his ability to distinguish between what was important and what was secondary. This created a distorted order of priorities, putting dealing with the knife at the top of the list rather than helping his wife clean her wound, which, of course, should have been his first priority.

Such serious cases are relatively infrequent, but milder failures are quite common. For example, take the considerable number of people who are always late. Despite sincerely wanting to, they never arrive on time for meetings and appointments, and can even end up missing their flights. In this case, the main item is the meeting, say, and of secondary importance is everything the chronic latecomer did before they should have left home – mop the floor, call to wish a friend happy birthday, tidy away the kids' shoes and all manner of other things that they simply had to take care of then rather than later.

A failure of this kind is something that can be turned around. Recognizing that a difficulty exists in distinguishing between the main event – the meeting – and the sideshow – mopping the floor – is the first step. A practical way to instil better habits is to reconstruct the events that preceded the meeting, check where the failure occurred and derive from this a lesson regarding how to approach future events.

Let's look at another example of a failure to navigate between major and minor issues in everyday life. It concerns a woman who enjoyed entertaining her friends every Saturday at four in the afternoon with tea and a particular cake, bought fresh from her local bakery 15 minutes before the gathering.

One Saturday, when she arrived at the bakery at the usual time, she was told that all those cakes had sold out, but she could find some at their other store that was not too far away. Without hesitation, she went there, bought the cake and returned home, half an hour late. Her friends were nowhere to be seen. That is because, in the meantime, they had arrived, waited in front of her locked door for quite a while, then had given up and gone their separate ways.

Such malfunctions, which result from a failure to distinguish between the main event (in this case, hosting the gathering) and the optional extra (the usual cake), are fairly common. If you search through your own memory, you will probably find that you are not immune to them either. However, those who have an excellent ability to make such distinctions are quite rare. People blessed with this skill often reach senior management positions, where they put it to use dealing with the countless situations that require decisions to be made.

A scientist who heads a research institute is a good example of such a person. In addition to her own research work, she must keep up to date with developments in her particular field and know what the hot research topics are, the pursuit of which boosts access to funding for the institute's research and the publication of results in prestigious journals.

Also, she must engage in teaching, be attentive to her colleagues, take care of administrative and financial problems at the institute, select the best people from among the many candidates who wish to join the faculty and many other tasks. Complying with all these demands naturally requires a high level of ability to focus on the essentials and prioritize tasks correctly.

Being able to distinguish between what is and isn't important is helpful to us all. It enables our executive function to fulfil another important cognitive role: decision making. To make the right decisions, we must choose the best option from among all the others before us. To complete the process, we need to stick to our chosen option and ignore those we discarded. Nonetheless, there is an important caveat to this last part, which is that the executive function must continue to monitor what is happening and, if environmental conditions change, it needs to get us to make another different decision, one better suited to the new situation.

It does not necessarily follow that people with highly developed analytical abilities will also be good decision makers. A personal experience of my late father's is relevant here. He was head of a department in a government institution. Once every few weeks, he told me, the boss would come in and present him with a problem that could be solved in different ways. The boss's analysis was always brilliant, but he was unable to reach any decision as to a solution. "Tell me, Chalamish," he would ask my father, "what would you do in my place?"

My father would consider the matter and give his opinion. The boss, much relieved, would thank him profusely and always accept his advice.

The excellent analytical thinking of my father's boss indicates that his mind had a sound infrastructure and a good workforce but, as they were without proper management, he was unable to do what was expected of someone in his position: make decisions on his own. Unfortunately, the existence of this gap in people's abilities is not something that employers are generally conscious of so, frequently, good employees are promoted to management positions without first testing their abilities in this area. The assumption is that a good employee will also be a good manager. Reality often proves this assumption to be wrong.

Managing emotions and behaviour

Another cognitive function that requires constant management is our emotions, aspects of which have been discussed earlier. Because of the great importance the brain attaches to emotions, it allows them to influence all brain activities, including motor, sensory and cognitive functions, as well as control of the organs in the body. That is why, when we are excited, our movements are fast and frantic; when we are sad, food loses its taste; when we are cheerful, it is easy for us to remember past cheerful events. It is also why anxiety makes your heart beat faster, even when the stress being experienced is second-hand, for example while you are watching a movie in a cinema or at home.

Poor management of our emotions affects and is reflected in our motor functions. For example, being highly stressed impairs both the planning and precision of our movements. Likewise, our senses are impaired, stress intensifying our sense of touch and diverting our gaze to potentially threatening stimuli. Cognitive deficits may be reflected in damage to someone's memory and ability to think, and stressful situations can weaken the immune system and cause malfunctions in the heart, lungs and other internal organs. But, usually, our management of our emotions runs smoothly, and for this we can thank evolution and our brain's understanding that living in relative harmony with others in a social framework is a necessary condition for survival of the human species. The same holds true for all animals that live in groups.

In the case of our ancient ancestors, the need to go hunting, with all its inherent dangers, fighting off predators and the forces of nature, intertribal struggles for territory and other hardships, required them to form partnerships with other humans. That is, to be social beings. To this end, the brain created a special factory (to refer to our original image for the brain) that gives us the ability to be social beings. The desire to be in the company of other people is not something to be taken for granted, so it is incumbent on the workers in this factory to create the right level of impetus for this in us. Failure to do so properly is recognized as a symptom of schizoid personality disorder, people with this disorder showing a lack of interest in social relationships or a tendency to avoid others. They also otherwise tend to isolate themselves and show emotional coldness.

Another task of the workers in our brain is to give us a deep understanding that each person around us has their own unique thoughts and desires that are different from ours, and enable us to consider what these other people are thinking. These special social skills are known as the theory of mind (TOM). There are people, including those with autism or schizophrenia, with brain damage that impairs these abilities.

One of the effective tests for checking the integrity of someone's TOM has been in use for several decades. In this classic test, the examiner presents to the participant and another person (the examiner's assistant), two sealed boxes – let's say one is blue and one is red.[4] The examiner opens the red box in full view of everyone and places a coin inside it. The examiner then closes the box and asks the participant where the coin is. The participant points to the correct box. The other person is also asked the same question and, after answering correctly, that person is asked to leave the room.

In the person's absence, the examiner removes the coin from the red box, places it in the blue box and again asks the participant to say where the coin is. After the participant points to the correct box, the other person is asked to return to the room and the examiner addresses the participant again, saying, "Now, if I ask our friend where the coin is, what do you think they will answer?"

The usual and correct answer, of course, will be, "In the red box." After all, the other person was not present when the switch was made. But if the participant suffers from a significant failure in TOM, it is likely that they will answer without hesitation, "In the blue box."

This is because the TOM-impaired person projects their thoughts onto the other person due to their lack of awareness that the other person is an individual, with other, different thoughts from themselves.

Failures in this area are not the exclusive domain of people with autism or schizophrenia. They may occur with lower intensity even in people whose TOM is considered normal. For example, consider times when you want to buy a gift for someone. It often happens that you choose an item that you like and are sure they'll like it as much as you do. The reality that the recipient of the gift may have tastes that are different from yours does not occur to you at all, though you are acting with the best of intentions.

Even if the workers in our brain do their job faithfully – they create the desire to be in the company of other people while also understanding that each of them is an individual with thoughts and desires different from our own – this is not enough in itself to always guarantee a good outcome. A necessary condition for achieving the goal, which is generating adequate closeness with others, is behaving appropriately in social situations. This task also falls to the executive function, which must ensure that we respond with behaviour that has been adjusted to suit a particular social event and the people who are there. In other words, it must organize for there to be the right facial expressions, body language, responses in conversations, silences and so on. All this can be summed up as behaviour that enables us to integrate into society.

For the sake of illustration, let us imagine an extreme example of a social situation. A man is sitting alone in a café. He is lonely and longs for company. Not far away sits a lively group of two or three couples. The man wants to fit in with them, so he walks up and, smiling from ear to ear, holds out his hand and says, "Hello, friends, my name is . . . I overheard you and it sounds really interesting . . . I would love to join you!"

It is not difficult to guess what the reaction of the people in the group will be. However, the behaviour of the man does not necessarily indicate deficiencies in the activity of the workers in his brain. It is also possible that he has a high IQ. The flaw lies in his behaviour manager.

There are people at the other end of the scale, whose executive function is operating at a high level of emotional intelligence. They are often defined as having personal charm. Yet, if we were to ask such a person how they are able to fit in easily with those around them, be accepted and popular, it's unlikely that they would be able to come up with an answer.

Attention deficit hyperactivity disorder

One of the commonest disorders of the executive function is attention deficit hyperactivity disorder (ADHD). Before we go into this in more detail, it is important to understand that it is not the result of a failure in the management of the brain. The executive function is acting properly but it is adopting a strategy that is at odds with what is considered normal or appropriate in society. Therefore, this phenomenon is called non-normative.

Attention and concentration activities are entrusted to a special area in the brain that allows us to be aware of the messages about stimuli that reach the cortex, both from the outside world via the senses and from within ourselves, such as a thought, internal pain, hunger and so on. Remember that, at any one moment, many messages relating to stimuli are received in the subcortex, but only a few of them go on to reach the cortex after filtering. At this stage, it is up to the executive function to decide whether, for the sake of promoting our survival, it is right for us to be aware of *all* these messages or just a few at that particular moment. Our level of attention is on a continuum: at one end, it is focused; at the other end, it is divided.

At the end of the continuum where our attention is divided, we will be aware simultaneously of a large number of messages, while at the other end, we will be aware of only one message out of the myriad messages that flowed into the cortex. This latter situation is what we know as concentration. Similar to the aperture of a camera lens, our ability to either focus in sharply on an object or open out to take in a panoramic image – our "attention aperture" (between the two extremes of focused and divided) – determines the number of messages we will be aware of at any given moment.

If I pose the question, "How open should the attention aperture be to best fulfil its role?", the seemingly logical answer would be, "Somewhere in the middle." As a matter of fact, it's much more complicated than that. So long as no unusual event is being experienced, it is indeed best for the aperture to be set somewhere close to the middle. But in everyday life, we encounter numerous events that require different aperture settings. For example, a doctor who runs the accident and emergency department at a hospital must be continually aware of the many things that are going on, so needs to be able to divide their attention. The doctor is at the extreme end of the continuum. The same is true for a supervisor working in a large garage in charge of several mechanics or a football player during a game or a nanny looking after several toddlers. In situations like these, the ability to be aware of many messages coming from the environment enables all of them, and us in our own life, to behave in the correct and appropriate way. However, we can picture other situations in which maximum focus and concentration is required. For example, the same doctor in the hospital while administering an infusion, the garage supervisor fitting tricky engine parts, the footballer about to take a penalty kick or the nanny while they feed one of their charges.

In most people, the attention aperture is adjusted to respond to the situation without us being aware of any change and without difficulty. However, for people with ADHD, this is not the case. The workers in the brain are doing their job properly – that is, they are able to operate with either focused or divided attention. The problem here is with the executive function, which, most of the time, organizes for the attention aperture to be set at the maximum division of attention end of the continuum. Only when there is strong motivation to concentrate on a particular stimulus does the aperture shift to maximum focus. To illustrate what this is like, we will follow the story of seven-year-old Omar, who has ADHD.

Sitting in class, Omar hears his teacher's voice and, at the same time, with the same intensity, he is aware of the sound of the wind rustling the leaves of the trees coming through the open window, of a fly hovering in the room, of thoughts about playing with his friend at lunchtime and countless other messages. If asked later what the lesson was about, he will probably shrug his shoulders in embarrassment. At the end of the school day, Omar rushes home to get to grips with a new jigsaw puzzle and, in a few hours, will manage to assemble all 10,000 pieces in a burst of continuous super-concentration.

Let's unpack what has happened here. Omar understands very well that he should listen to the teacher. He also knows that he will be tested on the subject of today's lesson. In a child without ADHD, these understandings are enough to make the executive function create maximum concentration on the teacher's words. Not so with Omar. If he is not highly motivated to devote himself to study, his executive function will leave his attention in a divided state. Therefore, as we saw, he attributed the same degree of importance to *all* the sensory and conceptual messages that reached him during the lesson. However, the challenge of mastering the thousands of puzzle pieces created strong motivation in him, which caused the executive function to direct his attention to the other extreme end of the continuum, to a state of maximum concentration.

ADHD tends to be hereditary, so it can be diagnosed at a young age. Evolutionary psychology explains its emergence and spread by positing that, in prehistoric times, it was a desirable way to be that helped to promote survival. Because the attention functions differently, such people were the best hunters, so they were considered desirable as mates and, thus, their genes spread widely and endured.

It is quite easy to understand why these hunters were so successful. At the beginning of the chase, with their divided attention at its peak, they would be able to detect signs of the sought-after prey better than most. Then, finding the animal's tracks would create a high level of internal motivation for them to dedicate themselves to following them, so their executive function would allow their attention to move to the other end of the continuum, with maximum focus being brought to the task of following the prey. The companions on the hunting trip would also try to locate the prey, but only the hunter with ADHD would manage to continually focus on a specific set of tracks while completely blocking out all the other messages, eventually guiding the hunting party to the animal's lair. At this point, everyone would aim their arrows at the mouth of the den, but only our ancestor with ADHD would be able to maintain that position, then aim without flinching, shoot accurately and win the respect of the tribe.

There has been a sharp increase in the number of children diagnosed with ADHD.[5] One explanation for this is that there is greater awareness of the condition among parents and teachers, which leads to it being spotted and investigated more often than before. Another explanation – a deeper one – comes from an acceptance of the relatively new concept of the flexible brain. To understand how this came about, we need to travel back in time, around 100 years, to a period when the lives of people carrying the genes of ADHD were quite miserable.

After surviving a rigid education system that meant they did not do very well academically, most people with ADHD had difficulty holding down the types of jobs that were then common in Europe and the USA – repetitive, monotonous work on factory production lines. These occupations did not require strong motivation but demanded long periods and high levels of concentration. As the brain's one and only goal is survival, it was forced to undergo a process of adapting to this reality. It did this by, among other things, changing the strategy of the executive function from continually dividing attention (provided there is no supreme motivation to concentrate) to behaviour that allows for concentration even when there is no strong motivation, only an environmental need for it. In this way, through most of the 20th century, ADHD was effectively submerged, so was much less visible than it is now that such ways of working have declined.

Fast forward to the present day and the trend has been reversed. In this technological, digital age, our brains are continually flooded with messages. A person who lives in this reality and copes simultaneously with many messages while keeping calm and functioning properly is held in high regard. This ability has even been rewarded with a respectable term for it, taken from the world of computers: multitasking. People with ADHD have a good chance of being included in this desirable category. An example of such a person is Oscar-winning director and screenwriter Alejandro Gonzalez Iñárritu (*21 Grams, Birdman, The Revenant*), who openly admits to having ADHD. Iñárritu's particular mental approach allows him to achieve super-concentration during the careful planning of scenes, filming and while editing the film, but be in a state of divided attention when he has to control the entirety of the production. The results speak for themselves.

From the above, it might be implied that for children diagnosed with ADHD, there is no reason to alter the strategy of their executive function. In fact, this is not the case because the education system has not yet caught up with the far-reaching changes that have taken place in modern society. Children are still expected to study within rigid frameworks (classrooms, tests, marks and grades) that require a high level of concentration during lessons, when doing homework and, of course, studying for and then sitting exams. In the past, people with ADHD had trouble adapting to these settings during their childhood.

But, later on, most of them levelled out to a greater or lesser extent, as a result of social pressures put on them after they left school. Today there is often the expectation that our children will continue their studies through secondary school and go on to university, so for those battling with attention and concentration in school, this situation will generally last longer than when more people left school as teenagers.

It can be said that in this context, modern society gives out mixed messages. Some cheer people with a nonconformist executive function for their ability to behave correctly in environments that keep it very busy, processing high volumes of messages, while others scold them for not being able to concentrate properly in situations that are considered significant in the eyes of the world.

How do we get out of this tangled mess? The generally accepted treatment is pharmaceutical, taking Ritalin, for example. The medicines prescribed are effective, in that they cause a shift from divided attention to concentration, but they are a far from perfect solution because the change is only temporary and there can be all manner of side effects. Besides, the gain of an improvement in the levels of concentration is accompanied by a weakening of the ability to divide attention. For example, during the lesson, young Omar will listen intently to the teacher's words but, during the break, will have difficulty playing football with his friends because this requires more divided attention.

As a result of adopting the idea of the flexible brain, brain researchers have come up with several natural ways to improve attention in people with ADHD.[6] One of these is to play an instrument in a group. This activity helps because it provides the opportunity to practise making rapid, natural transitions between concentrating and dividing attention – for example, playing a solo at some points, then playing in time with everyone else at others. Drumming is considered particularly effective in this respect as the drummer needs to alternate regularly from concentrating on one drum to playing several drums. Physical activities in a group setting, such as meditation and martial arts classes, are also considered effective therapeutic activities.

What all these options have in common is the making of quick transitions between concentrating and dividing attention. Practising this increases the ability of the executive function to make similar transitions in other situations in everyday life as well.

As is reflected in the name of this disorder, many children who have ADHD are also diagnosed as hyperactive. This presents as a particularly high level of physical activity and a tendency to relate in an excessive way to messages being received by the brain from the environment, so the children are easily distracted, impulsive, have difficulty concentrating and can be aggressive.

To understand the essence of hyperactivity, we must first recognize another aspect of the workings of the executive function, which is creating healthy curiosity. This is a kind of middle path between completely ignoring new messages brought into our awareness and paying too much attention to them.

What is regarded as a normal management strategy for the brain is to pay reasonable attention to messages perceived as important and ignore those perceived as unimportant. With hyperactivity, *every* message is perceived as important, so receives maximum attention. For example, the behaviour that would be expected of a child visiting the dentist would be that they would show interest in the variety of instruments and tools found there. Perhaps they would also ask questions about them. If they went that far, they might even ask permission to touch one of them. However, for a child with hyperactivity, they will not only have an uncontrollable urge to touch the instruments and press the pedals that raise and lower the chair but will also show an interest in the contents of the drawers and it may well be that they would give in to an uncontrollable urge to rummage through them.

As with ADHD as a whole, hyperactivity itself is considered to be a genetic trait that played a part in our prehistoric past, ensuring our survival, and has persisted to the present day.[7] It makes sense that paying maximum attention to all messages received from the senses allowed our hyperactive ancestors both to escape from dangers and to take advantage of opportunities to obtain food. When such hyperactivity occurred in people in tandem with disordered attention and concentration disorder, they would have been hailed as super-hunters. They would have enjoyed respect and prestige, and been seen as quite a catch, increasing the chances of them having children. Nowadays, moderate hyperactivity can advance our ability to look at messages received from the environment from an unconventional perspective, and even accelerate the way our brains process them.

The brain code as it relates to TOM is an important component of emotional intelligence. It is about the ability to assess the mental state (emotions, thoughts and behavioural intentions) of another person from a position of understanding that they have desires, feelings and thoughts which are different from our own. Sounds reasonable, doesn't it? Well, it is a mental ability and, as with all such abilities, there are some people who are very skilled in it. They do well in job interviews and on dates, and will always leave a positive impression.

Some tips for improving your emotional intelligence

Here are some exercises you can try to boost your social skills or TOM.

Method One

1 Choose someone dear to you and concentrate on getting inside their head – what are their ideals, dreams, goals, favourite things, activities, tastes, places . . . ?
2 Imagine you want to get them a gift. You want to find something symbolic and personal, that's not expensive, but will make them very happy when they receive it. What could it be? Consider their hobbies, preferences and past choices.
3 Give them the gift!

Method Two

1 During the day, focus for five minutes on a person engaged in some kind of work. It could be someone working in a café, a store or a colleague.
2 Watch that person, paying attention to their body language and the way they talk and behave.
3 Try to read their feelings and the thoughts that may be passing through their mind as they go about their work.

***

In this chapter, we have got to know the executive function of the brain and understand a bit more about how its administration of all the other functions can be improved. In chapter 14, all will be revealed as to how the brain code granted the brain's manager further wonderful, valuable abilities.

THE HEALING BRAIN

This chapter continues our exploration of the highest level of the brain's functioning. You will become familiar with your brain manager – your executive function – in a different role, a role for which metaphorical office wear is no longer appropriate. Instead, surgical scrubs will be better as we see how the executive function can treat our brain malfunctions in a targeted way. To do this, it still has its usual toolbox of resources to call on to fix things as we go about our daily life, but they are not enough for the purpose of healing. That is because correctly activating our neural networks alone will not repair damaged brain functions. To achieve long-lasting healing, structural changes are also needed.

From the top down

Here we are discussing an area that is indeed at the highest level as it is physically located in the highest part of the cortex, at the top of the head. The neural network found there is, in fact, our executive function and it has the power to promote or suppress all the brain's activities. The instructions issued by the executive function flow continually from that high point in the cortex down to the lower areas, so this activity is called "top down" (TD). These instructions must be focused and precise.

To clarify the concept of TD, I will describe an experiment known in brain research as "magnetic hands".[1] The participant is asked to extend their palms forward and up, holding them parallel and slightly apart. After doing this, they are told to concentrate on the idea that their hands are like magnetized metal plates, with opposite poles. Assuming that the participant has a clear mental picture of this idea, the researcher tells them that they will now be able to feel a magnetic attraction pulling these two plates together and, when they touch, they adhere to each other.

In many cases, participants *do* feel a magnetic pull and their palms come together without them consciously thinking to make them do this. It is clear to everyone that the participants' palms have no such magnetic attraction. The reason for the sensation and movement that occur is that two areas of the brain – one in charge of sensation for the hands and another responsible for the activation of the relevant muscles for the hands – received precise TD instructions from the executive function to perform the action in the format of two brain functions: concentration and anticipation.

Concentration

As we saw in chapter 13, the ability to concentrate, to be aware of only one message out of the multitude of messages received by the brain from the environment, requires an active closing of the attention aperture by our brain to reduce awareness of all the other messages. When we manage to focus on a certain thing or idea, the brain concludes that it is important and so is essential for our survival. Therefore, the executive function activates the relevant areas of the brain needed to process the messages relating to the object of our concentration, and sends operating instructions to the neural networks for various mental processes along the TD pathways.

Anticipation

In contrast to concentration, which is an ability shared by all animals that possess a frontal cortex (this includes many mammals), anticipation is a uniquely human ability. We have this wonderful extra thing we can do because we are the only species to have gone through the linguistic revolution, which, you will recall from chapter 6, is when some 70,000 years ago we began to use language to communicate, and this allows us to set expectations beyond the here and now. Expectation is a combination of emotion and concentration on an abstract idea mediated by language.

The magnetic hands experiment illustrates this well. From concentrating on the abstract idea that the hands are magnetized metal plates, and the resulting expectation of sensing magnetic attraction and the movement involved, the executive function surmised that this was the most important thing at that moment to ensure survival, so sent instructions out to the relevant areas of the brain to ensure it happened.

TD and healing

The executive function can perform operations in a TD format when the ideas we focus on, and our expectations, do not deviate from the routine activities of the brain that the executive function in office mode carries out all the time. For example, even if we concentrate on the idea that our arms have the capabilities of wings, and even if our expectation is that moving them will lift us off the ground, of course this will not happen. The simple reason for this is that we do not have the appropriate brain software for this because we do not have wings and so we have not lodged the brain activity necessary to move wings in our neural networks.

However, there is a grey area that science has not yet sufficiently explored, which consists of instances when the executive function manages to make things happen that seem to go beyond routine activity. I refer to significant improvements in a person's health, which can be achieved by using TD in a way that is not yet understood by science. Examples of this are when relief occurs in situations where the body attacks itself (autoimmune and allergic diseases), a significant benefit is experienced in cases of attention and concentration disorders, and even rare but documented cases of spontaneous recovery from cancer.[2]

Like every other complex brain activity, TD processing is also considered a form of intelligence. As with other types of intelligence, there are only a few people with TD abilities that are at a high level, on the verge of genius and, likewise, not many people with meagre TD abilities. The vast majority of us are somewhere in the middle. People who have exceptional TD abilities are able to implement almost any realistic idea by concentrating on it and anticipating it happening in reality. For example, if we ask such a person to stare at a blank TV screen while concentrating on a familiar movie with the anticipation of seeing it, they may well see the movie.

We can all *imagine* a movie in such a scenario, but only people with TD abilities at the top level will see it in a tangible way. As part of my postdoctoral work at the Weizmann Institute, I conducted an experiment using this very scenario with a woman whose TD processing abilities were highly developed. When the movie was over, I revealed to her that the TV had been off the whole time, but she was adamant that the movie had in fact been shown on the TV, it was that real for her.

In another experiment I conducted with someone who was similarly gifted, I asked him to concentrate on the idea that there was someone in an adjacent room who wanted to ask him a few personal questions.

I explained to him that both rooms were equipped with microphones and speakers, and I asked him to answer the questions that the other person asked in a loud, clear voice.

If I was being asked to do this, I would at most succeed in imagining that someone was talking to me, but I would certainly not hear their voice. That's not how it was for this participant, and I soon heard him say, in a loud, clear voice, "My name is . . . I'm a guy . . . I am single . . . my address is . . ."

Thanks to the flexibility of the brain, many of the disorders and other problems described in the previous chapters can be remedied. The use of TD abilities to correct malfunctions in the way the brain has been programmed, such as in relation to emotional, mental and behavioural regulation, does not require high-level abilities. Most people are capable of this. Indeed, in my opinion a high proportion of the population can benefit from putting the abilities they have to use. Furthermore, the flexibility of the brain allows us to improve our TD processing by means of exercises and training (see the end of this chapter for an example).

The use of TD processing for therapeutic purposes is not a recent development. Healers among our ancient ancestors have used these abilities effectively since the linguistic revolution, which allowed them to express in words ideas that might be beneficial to people who came seeking their help. Imagine the following scenario.

A member of the tribe suffers from severe headaches and seeks the help of a shaman, in whom he has complete trust. The shaman tells him to focus his gaze on his eyes, lays his hands on his head and says in an authoritative tone, "You can feel your headache fading . . . Walk around for a bit and the pain will disappear completely."

In many cases, this worked. When the shaman learned from his father how to conduct this kind of healing ceremony, he was unknowingly taking advantage of the patient's TD abilities. Concentrating on the idea the healer told him to, and expecting it to work – especially as it had worked previously for other members of the tribe – would achieve the goal of curing his headaches.

However, there is a fundamental problem with this form of treatment. Its effectiveness is limited because the TD processing paralyses the *expression* of the pain without creating a structural and functional change in the brain. As the source of the pain is not treated, the possibility of the headaches returning after a short while and in full force remains. For the treatment to be more effective, a more thorough alteration of the relevant brain software, the neural network, is required.

In the 21st century, it is the goal of most treatments based on TD processing to resolve the source of a problem. One such treatment is meditation.

Meditation

When a person is meditating, the activity in the cortex is slower than during normal wakefulness. Therefore, the amount of information that the cortex is able to process in this state is limited, making it easier for it to concentrate on one idea. As a result, it is possible to upgrade the ability of TD to move between focused and divided attention. An example would be moving from concentrating on breathing to concentrating on a sequence of thoughts that come to mind, regarding them like several passing clouds.

In this way, the brain can be trained to free itself from mental fixations and distressing emotions. There are different relaxation techniques that can be used to enter a state of meditation. One option is to be guided by the voice of a therapist, who will say something like, "In the next few moments, your body and mind will move into calmness, peace and pleasant rest. This feeling is getting stronger . . ." and so on. Responding to the therapist's voice implements the two essential components of TD processing: concentration and anticipation.

It is also possible to do self-guided meditation effectively, as in fact many do. In other cultures, meditation is achieved without the relaxation techniques usually used in the West. African and Native American cultures rely instead on repetitive body movements and dances accompanied by music, mainly drumming. Similar practices are familiar to Jews, who sway rhythmically while praying.

All the many different types of meditation are based on the same principle and originated in the Far East and Africa long ago, but meditation in the West is not a modern phenomenon. It was commonly practised in Europe until the Middle Ages, when the Church banned its use in religious ceremonies, so it died out in Europe but continued to flourish in Asia, Africa and the USA. Meditation later made a return to the West and has gained great popularity.

Brain researchers, too, have shown much interest in meditation as an effective, natural way to change aspects of the activity of the brain. Most of this attention has been directed at techniques based on Tibetan meditation, known as mindfulness or listening meditation. Also, the Dalai Lama dispatched many monks to prominent research institutes in the USA, where studies were conducted to see what happens to the brain during meditation and to examine its effects on the daily lives of the monks.[3] Among other things, the researchers discovered that people who meditate regularly display enhanced empathy, are more tranquil and their minds are endowed with better managerial skills than average. These conclusions prompted many neuroscientists to enrol in meditation workshops and see for themselves.

The importance of meditation and imagination as therapeutic tools employing TD processing has exceeded all expectations. Brain researchers have come to realize that imagination, which accompanies us throughout our daily lives, has far-reaching abilities to shape the brain. When we imagine something, we are both focused on it and able to develop an expectation of its fulfilment. This combination, when used correctly, can serve as an effective therapeutic tool. Ali's experience illustrates how helpful it can be.

Ali was in his 20s when he came to me for help. His appearance was striking, he was pleasant to talk to and had served in an elite military unit, so I was surprised to hear him say that he suffered from a lack of self-confidence.

"It's ruining my life," he said. "I am having trouble finding a suitable job, developing relationships with girlfriends or simply being sociable. I always go for the easiest option, which usually turns out to be the worst."

After listening in silence to his story, I asked him, "Try to remember, Ali, have you ever had a feeling of confidence in yourself?"

This was a key question because, to borrow terminology from computer science, TD applications are incapable of creating new brain programs; they can only run existing software. A negative answer from Ali would have indicated that his self-confidence software was missing, in which case, I would have referred him for more in-depth clinical psychological treatment than I could give him. Such treatment would check to see if he had any resistance to developing a sense of self-confidence, examine the essence of that resistance and attempt to overcome it. Naturally, that would involve a long course of treatment.

Ali thought briefly, then his face lit up and he said, "I like surfing. When I'm riding a surfboard, I feel on top of the world. At any other given moment, I feel really low, without a bit of confidence in myself." I immediately thought that here was something we could work with, so I asked him to close his eyes and picture the scene I was about to relate.

In detail described to Ali a sunny day at the beach. There was a pleasant breeze and waves breaking perfectly for surfing. He was walking toward the sea. His feet felt the warm sand and he walked into the sea with his board. In came a great wave.

"You jump on the surfboard," I continued, "you're on your feet and the wave is carrying you . . ." At this point, I fell silent and watched him. From the expression on his face, I concluded that he was focused on this idea. The smile that played on his lips testified to the good feeling it gave him.

"Tell me, Ali," I asked him, "how do you feel now?"

"On top of the world!" was the answer that came back, no hesitation.

"You didn't surf any waves just now," I said, bringing him back to reality, "you were here in this room, yet you felt confident."

Ali was thrilled. "Nothing like this has ever happened to me in my life," he murmured, still smiling.

Consequently, our first meeting was a great success and, while it is true that one session did not solve Ali's problem, it did send an initial message, directed to his self-confidence software, ordering it to be more available to him.

In the meetings that followed, we returned to this imaginary scenario, and I instructed Ali to practise it himself as well. It was a learning process in which imagination was combined with a strong motivation to "get out of the rut", as Ali called it. Soon his self-confidence software began to function well and, increasingly naturally, it became part of his daily routine. This was evidence that his brain was learning and new connections were being formed between neurons until it would no longer heed the messages that formerly activated his feelings of a lack of confidence and sadness.

The role of reason in healing

As we have seen in previous chapters, emotion is supreme and usually dictates our perception of reality. It seems that this was the brain's only option for our ancient ancestors until the linguistic revolution, when language enabled humans to enlarge their picture of the world around them. However, this also created the possibility that reason could dictate not only our perception of reality but also our emotions. I will illustrate this by telling you another story.

Abbie expresses her deep love for Tom and her strong desire to share her life with him. He is beaming with happiness. "It's also my wish," he replies. "But I have to tell you something. I went for a test and I am HIV positive."

You don't need to have a particularly vivid imagination to realize that this piece of news will cause a deep change in Abbie's emotions. Let's examine what has happened. Her reality did not change but, as a result of Tom's news, a new piece of knowledge was added to her reality and that new knowledge caused the change in her emotions. This change was made possible by the developments that followed the language revolution. After all, before that, we could not express and pass on such information in the here and now. The ability to influence emotion by seeing reality in a different light by means of language forms the basis for cognitive behavioural therapy (CBT), which uses TD processing. Brain researchers consider CBT to be an effective and practical tool.[4]

Cognitive behavioural therapy

When the brain interprets a certain situation as threatening our survival, it creates a feeling of suffering and fear, which we should interpret as a kind of alarm. The goal of CBT is to create a new perception of reality (reality itself cannot be changed, of course), in which the person will conclude that they are not, in fact, in a situation that threatens their survival. When the brain no longer perceives reality as threatening, it will silence the alarm, which is the feelings of suffering, anxiety, fear. Thus, while it is usually emotion that drives us, CBT aims to shift the reins from emotion to rational thought.

This task is assigned to the TD processing system, and the two conditions for its success are that a person concentrates on the therapist's words and has a high level of expectation that putting the CBT strategies into practice will have a positive effect. I will demonstrate the method used for CBT by relating how I used it to help Kenji.

Kenji, a young man working in IT, opened up to me and explained that recently he had been experiencing distress so severe it was disrupting his life. He was anxious, depressed, unable to concentrate at work, lacking appetite and suffering from insomnia. I asked him if he knew the reason for this. His answer was unequivocal: two months previously, a new supervisor had been appointed in his department.

Kenji explained, "The previous manager valued my professional skills highly and often praised my creativity. She also promoted me. The new manager is the complete opposite of her – he's a guy in his 50s, very set in his ways and impatient. From the very beginning, he treated me with disdain, talked down to me and criticized everything I do. He attacks me for no reason and embarrasses me in front of everyone else."

I listened attentively to Kenji and, while doing so, I had an idea. But, for it to succeed, his focus would need to be razor-sharp.

I began, "Concentrate hard on what I tell you next. This tough boss has arrived at a new workplace. As you said, he's already in his 50s and, I don't need to tell you, in IT that's considered old, so he likely fears for his future and continually feels like his job is at risk. His goal is to get rid of anyone who might be his replacement, and he probably realized very quickly that you are the natural candidate. That's why he's antagonistic toward you."

Kenji pondered on what I'd said for a minute or two, then broke the silence and let out a sigh of relief. "It all makes so much sense", he said.

In our next three meetings, we expanded on this idea and I could see that Kenji's understanding was getting stronger, which showed the deep and meaningful learning he had achieved. At our fifth meeting, which was also the last, he told me, "Even today my boss was criticizing my work in a negative way, but the more he got carried away, the more I was smiling to myself."

The CBT had achieved its goal. Although the reality remained unchanged and the boss carried on as before, Kenji's perception of that reality underwent a fundamental change. Kenji no longer saw himself as threatened, so his brain switched off the alarm and put an end to his mental suffering.

Hypnosis

Hypnosis is still seen by many as, at best, a mysterious magic act and, at worst, one of those old stage shows put on by disreputable scammers who faked or exploited the brain's abilities for entertainment purposes. This image has been reinforced by movies such as *The Manchurian Candidate* (1962) and other psychological thrillers, but this needn't be the case.

In truth, hypnosis is a legitimate, efficient form of therapy that makes use of TD processing, though brain researchers did not properly understand this until the beginning of the 21st century. Following major breakthroughs in everything related to TD, it can be said that hypnosis has since been upgraded to a legitimate form of treatment involving the imagination and CBT.

During hypnosis, you are in a state similar to that achieved in meditation, in which, as we saw earlier in this chapter, the activity in the cortex occurs more slowly than usual. As it is not possible to concentrate on several subjects simultaneously in this state, your attention is focused on the words of the hypnotist or therapist. This increase in concentration, as well as the anticipation that the hypnotist's words will be applied to the difficulty to resolve it, means that the TD processing reaches maximum effectiveness. This is why hypnosis is considered one of the most effective methods of therapy.

Unfortunately, some baseless myths still persist that hinder hypnosis from being adopted more widely than it has been so far. One is that, under hypnosis, you cannot exert your own free will, and you become a slave of the hypnotist. This is not true: a hypnotist cannot have any kind of omnipotent influence over those they have hypnotized.

To understand why this is the case, it is important to remember that the TD process is initiated and takes place in the brain of the person undergoing hypnosis, not the hypnotist's. Therefore, if the hypnotist's ideas are contrary to what the hypnotized person perceives, they will refuse to concentrate on them, so the hypnotist will have no influence. For example, say the hypnotist orders the hypnotized person to jump out the window, they will not cooperate, unless they were having suicidal thoughts of their own in the first place. The hypnotist's words will also have no effect if the hypnotized person is not concentrating on them but is thinking about something else instead. This will be the result as well if the ideas the hypnotist is suggesting go beyond the brain's usual capabilities, such as an instruction to fly around the room.

Another falsehood is that a person may not wake up from hypnosis. The word "hypnosis" derives from the name of the Greek god of sleep, Hypnos.[5] However, hypnosis is not a state of sleep; it is one of meditation. During my clinical work using hypnosis as a therapy, I have had patients fall asleep during a session, but this was due to fatigue, not the hypnosis process itself. When I noticed this happening, I knew that everything I said to them from that moment on would be in vain because, during sleep, the ability to listen is obviously greatly diminished, so my words would not reach them. So, I would gently wake them, suggest that they wash their face and then continue the treatment or arrange for them to come back when they were less tired.

In Israel, where I practise, a law was passed in 1984 restricting hypnosis to clinical psychologists, general practitioners and dentists. The permission to engage in hypnosis was later expanded by the Ministry of Health to include medical and psychological diagnosis and treatment, scientific research, and to refresh the memory during police investigations. In South Africa in 1997, an amendment was made to its Health Professions Act, limiting hypnosis and hypnotherapy to licensed psychologists and mental health practitioners. In the UK, there is the Hypnotism Act 1952, but that concerns only stage hypnosis and does not relate to therapeutic uses of hypnotism at all. Otherwise, and elsewhere in the world, its use seems to be largely self-regulated rather than subject to laws.

The impetus for the legislation in Israel was a shameful case that came about during a stage show in 1975.[6] The hypnotist located a handful of people in the audience with highly developed TD processing abilities and invited them on stage. One of them was a 16-year-old girl, Yaffa Svisa. The hypnotist had no way of knowing, but Yaffa belonged to a group of people whose brains use TD in an abnormal manner, which is that whenever they feel acute distress, their brain causes paralysis and physical pain, and disrupts messages from the senses. The name given to this phenomenon is dissociative disorder.

During the show, the hypnotist laid Yaffa across three chairs, then removed the middle chair and told her to harden the muscles of her torso. Then he sat on her. The laughter from the audience died down when it became clear that the hypnotist was unable to communicate with Yaffa and end the hypnotic state.

The scientific explanation for this is that the awkward situation activated a threat circuit in Yaffa's brain that paralysed the muscles in her limbs, vocal chords and eyelids. The unfortunate girl was hospitalized but, after a few days, her condition was unchanged.

Yaffa's case was referred to Maurice Kleinhaus, a psychiatrist who is considered the father of hypnosis in Israel. With the help of relaxation techniques to instil calmness, self-control, together with confidence in and cooperation with him, he was able to reactivate Yaffa's inhibited functions.

Kleinhaus told Yaffa his name and explained that he was a much better hypnotist than the one who put her in this situation. He asked her to blink once if she heard him, twice if she didn't.

She blinked and that was the beginning of the therapeutic relationship.[7] After a while, Kleinhaus and Yaffa walked out to meet her worried parents. Yaffa had returned to herself.

Along with many of my peers who are allowed to treat people using hypnosis, I also believe that there is a place for the well-regulated study of hypnosis used in clinical situations, but there is no need to limit its therapeutic use in such a rigid manner as is currently the case in Israel.

Biofeedback

Biofeedback is a relatively new treatment method that combines sophisticated technology with brain research. As noted in chapter 8, electronic devices are used that are capable of continually and accurately measuring small changes in your physiological functions, such as brain waves (also known as neurofeedback), muscle tension, heart rate, blood pressure, emotional arousal, body temperature and respiration, so you can develop better control over them. The results of this continual measurement are the feedback, displayed in real time on the device. When you observe a deviation from an acceptable level of, for example, your heart rate, by combining your strong motivation to lower it and concentrating on making it happen, you will usually achieve your desired goal. Thus, your executive function is activated via TD processing to send messages to the area of the brain related to heart rate control to bring it down to a calmer level. However, once you disconnect from the device, your heart rate will likely climb back to its previous high level because, in this initial stage, the learning you have acquired in this process has not yet been made permanent. In other words, no long-term change has occurred in the relevant neural network. But as the treatment progresses, by repeating this process, the necessary changes to the neural network will occur.

To clarify this a little, let me remind you about Ali, the young and insecure surfer I introduced earlier in this chapter. Even in Ali's case, his situation returned to how it had been after our first meeting, but once he had a few more sessions and repeated the process I taught him at home, he was able to make the improvement permanent and gain in self-confidence. Therefore, biofeedback therapy by itself cannot solve problems. It relies on people being highly motivated to get their heart rate or other health concern under control, as it is their persistence that will create the path to learning and, eventually, the process required to make the improvement will then be automatic.

The number of biofeedback institutes around the world is increasing, and the technology used in the devices is continually being improved, so this form of treatment is becoming more and more effective.

Placebos

A placebo is a medical procedure that does not involve administering any medicines, and a placebo effect is the effect that such a procedure has. When people use the term placebo, they usually mean a dummy medicine, such as a pill containing only sugar, that the person taking it believes to be a medicine that is supposed to have an effect. But a placebo can also be an ultrasound device that hasn't been turned on, but the therapist has told a person that it will make their headache go away. It's interesting to note that some studies have found the placebo effect exists even when the people involved know that they've been given a placebo. Ted Kaptchuk, from the Harvard Medical School, prescribed harmless sugar pills to people suffering from irritable bowel syndrome (IBS), and told them that's all they were, yet 59 per cent of the patients experienced an improvement in their condition.[8]

Many of the treatments that were accepted in antiquity relied heavily on the placebo effect. In modern times, placebos are deliberately used in trials of new medicines. Half the participants in these experiments receive the new medicine being tested, while the others – the control group – receive a dummy drug that contains no active substances. None of the people taking part knows which medicine they've been given. So comparing the results for the two groups makes it possible to assess the true effectiveness of the new medicine without any of the psychological effects of bias that would likely occur if the participants knew what medicine they were taking. Due to the aforementioned placebo effect, it's expected that some of the control group will display positive effects anyway. That is, a proportion of those who receive the placebo will report relief, even curing, of their symptoms. For a particular drug to be approved, it must perform better than the dummy pill.

Neuroscientists have been trying to convince the medical establishment that placebos can be used not only in trials for new medicines but also as effective means of treatment. Their position is based on both knowledge of the mechanisms activated when TD processing is applied and the results from studies that testify to the power of placebos. For example, consider the case of a number of people who experienced inflammation of the knee.[9] About half of those who underwent surgery to treat the inflammation reported significant pain relief. Unknowingly, some patients did not have the full surgery; it only looked like they had. They were anaesthetized, incisions were made and then sewn up and a similar amount of pain relief medication was given to them as was given to those who had the operation. The doctor concluded that if the surgery is no more effective than a placebo, then it's time to cancel this type of surgery completely.

Unfortunately, the doctor did not emphasize an equally important fact, which is that the surgery *is* effective, but the power of the *expectation* that the patient would improve or the placebo effect is just as good. For this reason, neuroscientists would inevitably suggest that orthopaedists should promote treatments based on the placebo effect for patients suffering from inflammation in the knee area. There is corroborating evidence supporting this view from various experiments conducted by neuroscientists such as Fabrizio Benedetti and others.[10] I remember, at a conference in 2012, Benedetti described the following research work he was conducting with colleagues.

Consent was obtained from burns patients and their doctors for the patients to become active partners in research related to the placebo effect. It's important to understand that severe burns over large areas of skin are one of the most painful types of injury you can have. For pain relief, infusions containing morphine were administered three times a day at the same hospital. During the first phase of the study, the patients underwent a functional magnetic resonance imaging (fMRI) scan while under the influence of the morphine. As expected, the scans showed a decrease in activity in the neural network relating to the pain.

In the next phase of the experiment, an inert saline solution (0.9 per cent NaCl) – the placebo – was substituted for the midday morphine infusion without the patients or the nursing staff being notified of this switch. As expected, the patients did not notice the change and did not complain about pain in the afternoon.

Not content with this outcome, Benedetti said that he and the team also continued the fMRI scans while the patients were under the influence of the saline solution. The results of the scans were the same as when the patients received the morphine infusions – that is, there was a reduction in activity in the neural network relating to the pain.

This was, of course, the placebo effect. The patients expected significant pain relief because they assumed that the transparent liquid flowing into their veins was their new best friend – morphine.

Later in the experiment, Benedetti said that he and the team met with the patients and obtained their consent to substitute a saline solution for the morphine the next day. The nurses who cared for the burns patients were there, too, and accepted that this was what would happen. But this time the patients were not given a saline solution. Instead, the doctors were instructed to give the patients morphine infusions. Thus, when the morphine was administered, both the patients and the nurses thought that they were receiving a saline solution.

Benedetti went on to say that what followed was not surprising. The patients, who believed they had received an inert liquid, cried out in pain, even though they had, in fact, received their usual dose of morphine.

A conclusion reached by Bendetti as a result of his work is that a significant part of the beneficial effect of *all* medicines is due to the placebo effect. Because of this, even the most effective medicine will not yield its full potential benefit if the doctor who prescribes it does not create in the patient an honest expectation that their condition will improve. This is a crucial lesson for doctors who neglect to make eye contact with their patients when they are sitting in front of them, instead tapping away on their keyboard, being impatient, distracted, interrupting or showing a lack of interest.

Expectation, as one of the central elements in TD processing, is also an effective tool when it comes to bringing up and educating children. Parents and teachers who honestly and transparently expect their children and students to succeed in school and exhibit good standards of behaviour socially, and convey messages that are in line with those expectations, go a long way toward them being fulfilled. This understanding, it is important to emphasize, is based on scientific foundations. Expectations, like prophecies, tend to fulfil themselves.

Some tips for improving your TD processing to increase your wellbeing

As you now know, the brain code is programmed to promote our survival and wellbeing. To put it to work in a positive way in your everyday life, you need to embed clear and realistic messages in your brain. After receiving your messages via TD processing, your executive function will be able to direct the functions of attention, memory, sorting the wheat from the chaff, and decision-making to make the best scenario happen because it will have received the message that it needs to do this to ensure your survival.

Here is one way to make this happen.

1 Pick up a pen and make a list of what you wish for yourself over the coming six months. Divide your list into several categories, such as romance, family, employment, health and social matters.
2 For each category, write one or two sentences that express what you wish for most for that item in that category.

3 Be very clear – your executive function needs to be able to understand!
4 When you have finished writing, read your list aloud, concentrating on each item.
5 Your messages are now encoded in your brain. Fold your list, store it away and go about your daily life.
6 Six months later, read your list again and see how much you have changed things for the better. In the interim, your executive function may have led you to crossroads that without having completed this TD processing enhancement exercise, you would not have been brought to. By means of this intentional management, your executive function can manipulate your preferences so that they are consistent with the messages you have written down.

***

Improving our TD processing abilities by enhancing the flow of messages from our executive function at the top of our brain down to the neural networks for lower-order areas, and from there to the rest of the body, can work wonders. Whether it is imagination therapy or hypnosis, biofeedback or the placebo effect, in severe cases, such an intervention can effect a transition from a life of suffering to a tolerable and even good life. In mild cases, although the change is much smaller, that 10 per cent difference, for example, can transform a difficult existence into one that is much more satisfying. Whatever the size of the improvement, give it a go – it is sure to be worth it.

CHAPTER 15
WITH LOVE IN MIND

You are probably familiar with those times when some innocent small talk with someone has a profound effect on you. Well, it happened to me when a friend told me about his relationship with his wife.

FRIEND: The love I feel for her reminds me of the days of my youth. Back then, I was helping my father renovate houses and I often had to hold up the top of the frame while we were replacing the front door of a house. For the longest time, I would support that top crossbar with both hands but, after a while, I could feel my arms giving out, so I would tell my dad that I couldn't do it for much longer and he would stop the work for a moment, so I could have a rest. But everything changed in a flash when I realized that the best way to hold up the top of the frame without getting tired would be to use my hands alternately. At any moment, one hand could support the frame while the other rested. Then, when the working hand got tired, I could swap it for the other, and so on. I tested my idea on the job and it worked perfectly – I was able to hold the frame in position steadily for a long time.

ME: Interesting story, but how does that relate to your relationship with your wife?

FRIEND: Well, she's the love of my life and, usually, I feel very passionately about her. But, what can I do, there are times in life when the flame of passion burns out. And when that happens, I keep our love alive by using my other arm. That is to say, I put emphasis on the wonderful friendship and togetherness that exists between us. We have a pleasant time talking, we might sing or play music together, and, sure enough, the passion is rekindled.

I can recall the excitement that gripped me while I was listening to my friend telling me this. I could visualize the harmonious brain activity associated with his relationship and I rejoiced for him. We will come back to my friend and his wonderful story later in this chapter.

Next, we will come to understand the essence of love, as illuminated by the insights arising from brain research, why it is so connected to our physical and mental health, and how we can promote it in our daily lives.

What is love?

Love is defined as a group of basic and higher-order emotions associated with a feeling of intense affection, profound unity and an intense creative drive. Love can be felt toward a person (ourselves or others), an animal, an object, a place or even an idea, and it is characterized by a search for intimacy with the beloved.

Its intensity is located on a continuum between two opposite poles: indifference toward the beloved and excessive love, bordering on obsession or addiction. Thus, the power of love can be expressed by locating it on a scale somewhere between these two extremes.

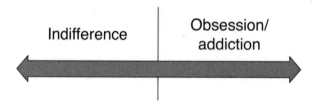

To illustrate this, let's do a little exercise. Place the intensity of love you feel for each of the following somewhere along the above scale.

- yourself
- your partner (current or past)
- your good friend
- your pet
- a car
- an unforgettable movie
- the regulation that requires wheelchair access in places of entertainment

Pay attention to the changing intensities of love you feel for each of these items. Like so many brain functions, love is on a continuum.

Where does love come from?

The human brain generates the feeling of love and, like our other brain functions, the precursor for this is the activation of a neural network, this time relating to motivation. Only when this network starts to play its role will the emotions and accompanying impulses expressed as the kind of brain activity we call love get the green light. It is clear that motivation is necessary for this to happen from comments we often hear, such as, "It's strange, he ticks all the right boxes – he's such a nice person, he's good looking, with an apartment and a steady job. I want to love him but I simply don't." This person's brain has decided not to activate the motivational neural network for love.

Another everyday scenario with a similar outcome is where a man goes to buy a suit for an event. He tries one on and those who've gone with him compliment him on how good it looks, the cut, the quality of the cloth and the colour, but he does not connect with what he sees and hears. He simply can't manage to like this suit. His brain has concluded (which may change later) not to activate the motivational neural network for attachment to the suit right now. A well-known phrase people say in such situations when they have not been convinced to change their emotional and thinking patterns is, "Please, don't confuse me with the facts."

As mentioned in chapter 1, our brain consists of two independent parts. There is a deep brain area called the subcortex, which detects threats and opportunities, then processes them with a focus on our immediate survival. The other part is an outer brain area called the cortex, which is concerned with our long-term survival and wellbeing. Yet, interestingly, the brain function that relates to expressing love can arise from both parts of the brain – the subcortex and the cortex – one at a time, both at the same time but separately, or both in unison. Therefore, in my own case, I love many things – my home, my mother, my friends, my car, my dog . . . I can also feel a general love for everyone, but the love I feel for my daughter is deep, so the source of my love for her is from both my subcortex and my cortex. However, the love I feel for a friend or for my dog springs only from the cortex. So, while I feel true love for all the aforementioned people, animals and things, there are fundamental differences in my emotional and behavioural expressions toward those objects of my love. That is because two important parameters characterize love: the part of the brain it originates from (the cortex or subcortex) and its intensity.

Can love at first sight last?

As noted, the key condition for love to begin in the subcortex is that, at the outset, the neural network for motivation is activated. This preliminary activity will confirm that the object of our love – be it living or inanimate – might promote the immediate survival of the person feeling the love. To this end, under the radar of our awareness and cloaked in secrecy, discussions take place in the brain that lead to a decision as to whether to generate motivation and enable love to be sparked into life or not.

Highly sophisticated chemical laboratories in the brain examine the messages coming from the senses during initial interactions with the potential focus of our love. An innocent conversation, a hug or a kiss on the cheek were designed, among other things, to collect data and present it to the deep brain for analysis. With the help of our senses of smell, sight, taste and touch, the data needed are collected and, after a chemical/electrical test, a decision is made. Does the person or the object in front of us contribute to promoting our survival or not? If the answer is "Yes", a "Go" signal will be given to activate the neural network for motivation in the subcortex and flip the switch to the "love" setting. If the answer is "No", the neural network for motivation will not be fired up and no messages for love to get underway will be generated and sent by the subcortex.

Two important items characterize these reactions of the neural network regarding love. One is the intensity of the motivation generated. This can be positioned along a scale, and the laboratory in the subcortex that reaches the "Yes" decision will determine whether the intensity of the love we feel will be low, medium or high. We have all heard of love at first sight. This would be a decisive and quick answer on analysing the results of the sensory messages. We usually only talk about this kind of love in a romantic context, but it can also be expressed in other ways. Such an instant, positive reaction is also what happens when an interviewer realizes very quickly that the person before them is the candidate who is best suited to the job, or it can be directed toward some clothing or a special item in a store window, a picture in a gallery, a potential pet or a new friendship.

The second important characteristic of the neural network's reaction is its instability. Love will always be conditional. The laboratory will continue to assess messages it receives after making its decision – to see whether the object of our love still promotes our survival – and will adjust the intensity of the motivation for that love depending on the latest results. That is why the passion you had for that pair of shoes may fade over time. The fitness equipment you emptied your bank account to buy might end up gathering dust. People who were madly in love break up.

Love that originates from the subcortex will only be stable if there are processes that ensure the brain learns, as this perpetuates its continuation in everyday life. When learning happens, love will continue to be generated without any need to repeatedly activate the neural network for motivation. Evidence that learning has occurred is the forming of new connections in the neural network, and these new elements in this, the brain's software, joining them up to and establishing them as part of the routine activity of the brain.

The capacity for love from the subcortex to become stable is related to a characteristic of the brain code called temperament, which in turn relates to the genetic basis of our personality (see chapter 9). To understand this a little better, here is the story of Will, who experienced extreme emotional instability.

The outcome of this instability was that Will had severe difficulty in maintaining social relationships and holding down a job. He would experience sharp emotional transitions, so enthusiasm for a potential partner would, a few days later, turn into feelings of indifference toward them, and feelings of energy and motivation about a new workplace would quickly turn into dissatisfaction. Similarly, an item in a store might captivate him but, a few days after buying it, his interest in it would evaporate.

Despite all this, Will did have one long-lasting, steady love. It was love that had originated in his subcortex, but later became established in his cortex as well. It was love at first sight when he set eyes on a particular horse. He visited the mare every day at the stables, groomed her, cared for her and never got angry with her, even when she misbehaved.

The message from Will's story is extremely important. Our knowledge about the brain is still far from complete, so researchers and doctors should be careful to avoid making rigid diagnoses when someone's condition is complex. In Will's case, I advised him to assist a child with a learning disability who might benefit from riding lessons. All this, of course, took place with the supervision of a qualified counsellor. I suggested this on the basis of my deducing that his brain's software for emotional stability – his love for the horse – might be used to expand the emotional stability he did have to the rest of his life. It was a success! Will became attached to the child and was able to be emotionally stable throughout the riding lessons.

From Will's experience, we can see how love originating from the subcortex can be developed when there is a deep understanding by the brain that the object of the love has the ability to promote our survival. The opposite can also happen. There can be a dramatic decision to resist any relationship with the object of our love. Then, the activation of the neural network for motivation will be suppressed and the resulting emotion felt will be one of indifference.

Love that lasts

Let us now consider how love can also originate from our cortex. When we feel love for a person, pet, object or idea, but the reason for this love and the brain activity behind it has happened in the cortex, that feeling will be completely different from love that was generated in the subcortex.

The cortex, as you now know, is the grey matter that makes up the outer part of the brain and works independently from the subcortex. Like the subcortex, though, the cortex receives a constant stream of messages from the senses to process, but its goal is to promote our survival in the long term, which means that it works toward ensuring our wellbeing, whereas the subcortex focuses on the immediate future. Also like the subcortex, the cortex has a highly sophisticated chemical laboratory, which receives and processes messages about the environment. Once again, the output of this processing will activate the neural network for motivation, which in turn will generate feelings of love. But, remember, this time the source of the feelings of love is the cortex. So what does the love laboratory in the cortex do that makes the feelings of love it generates different from those of the subcortex?

The answer is that the laboratory in the cortex seeks to determine whether the person, object or idea that you perceive as a result of information picked up by your senses will promote your long-term wellbeing. If the answer is "Yes", the neural network for motivation in the cortex will be activated and love will begin. Assuming that it is reciprocated, that warm and pleasant rush we feel will urge us to seek greater closeness, making the "Yes" answer more powerful, and if the feelings are more powerful, the desire for intimacy will follow. If, instead, the laboratory's answer is "No", then the neural network for motivation will not come to life and there will be no feelings of love.

When the cortex does give a "Yes" answer, it does not end there. After a trial period – during which the object of our love is obliged to keep passing the laboratory's tests to gain the trust of the neural network for motivation – comes the learning phase, with its sought-after outcome of stability. To enjoy a long lifespan in the cortex, love must balance two parameters: who to love and how much to love. Regarding who to love, the results from studies support the notion that it is worth loving people who are similar to us in terms of their background and character traits.[1] When considering how much to love (not too little and not too much), it is crucial that the relative strengths of the feelings generated by both the cortex and the subcortex are balanced. An excess of love from one or the other may lead to, for example, obsession for love from the cortex, or addiction for love from the subcortex, as the following examples demonstrate.

Hoarding is a great example of a feeling located at the obsessive end of the scale for attachment that has originated in the cortex. Hoarders become attached to items that most of us throw out or recycle, such as old papers or clothes. When asked, "Why do you keep all this junk?", they say things such as, "You never know when you might need it." The motivation for this behaviour is a misplaced attachment, emanating from excessive meaning being linked to these items.

Extreme greed has its root in a similar mechanism, but is an example of an addiction that has its origins in the subcortex. In this case, the pleasure brought by money and possessions tips into an excessive love for these things. As with all addiction, it leads to a greedy person always wanting more and never having enough.

The love that parents usually feel for their children mainly comes from the cortex. This component of parental love provides support for parents in moments when they experience threats to their survival that the challenges of having a child pose, such as sleepless nights, tantrums and financial stresses. Sometimes this love may fade temporarily and then an unpleasant, unfamiliar and frightening emotion of anger, rejection or even indifference toward those closest to us may be experienced.

Love may also disappear for extended periods of time. For example, during an episode of clinical depression brain activity in general is suppressed and the neural networks in the cortex that generate feelings of love are also affected. I saw this happen during my psychiatry internship. An older woman was in my care. Her grandchildren were her entire world – she collected them from school and kindergarten, fed them lunch, helped them with their homework and played with them. They had a wonderful relationship. Nonetheless, she was stricken with severe clinical depression requiring hospitalization and, during this time, she showed indifference toward her grandchildren.

As we saw in chapter 5, clinical depression tends to come in waves and, in the vast majority of cases, a wave will pass after a few months. Indeed, the grandmother recovered from her depression and then the strength and vitality of her love for her grandchildren returned. However, you do not need to go through an episode of depression to experience a temporary decrease in your feelings of love that come from your cortex. During an outburst of anger, the vigorous activity going on in the subcortex takes precedence over the influence of the cortex, creating a temporary pause to its usual process of generating feelings of love. A friend told me of an incident that clearly exemplifies.

A storekeeper called my friend to let him know that his 14-year-old son had been caught stealing some chocolates from his store. My friend thanked the man for deciding not to report his son to the police and promised to take care of the matter.

"You understand, Yossi," my friend said, "I truly love my son but, after the call from the storekeeper, I felt enormous anger and great disappointment. In those moments, my deep love for him was erased. It was frightening but, fortunately, it didn't last long. After he apologized to me and the storekeeper, my love for him returned. I had calmed down."

Love and culture

Let's now take a step back and look at the forming of our feelings of love from a different perspective. These feelings, when they have their origin in the cortex, occur when there is an understanding at the deep layer in the brain that the object of our love will advance our wellbeing and contribute to the promotion of our happiness. But what is it that contributes to our wellbeing? What brings us happiness?

The parameters we set that enable us to make decisions in answer to these questions are based on the cultural values we grew up with. An important concept that has arisen in brain research relating to this idea is of culture shaping the brain. That is, the cultural influences that have the power to engrave in the cortex concepts such as ideals for appearances, character traits and other notions that we believe will contribute to our wellbeing.[2]

Oxford Dictionaries announced "post-truth" as the Word of the Year in 2016, which means that the minds of two individuals with different cultural origins will be so different that they interpret whether a certain character trait will promote their wellbeing or harm it in contrasting ways.[3] For example, one person may interpret traits such as gentleness and being softly spoken as positive. Therefore people who exhibit these characteristics will spark their neural network for motivation in their cortex to create feelings of wanting to bond with such people. But another person, from a different cultural environment, may interpret these traits completely differently, for example, as signs of weakness. Their cortex will give such people a low score in terms of contributing to their wellbeing, creating a feeling of aversion to gentle people. Who is right and who is wrong? In an era when truth is relative, there is no clear answer; each of us can have our own truth can change over time.[4]

Head or heart?

So far, we have separated love that has originated in the cortex from love that has begun in the subcortex but, in reality, most situations will combine feelings of love from both parts of the brain in varying intensities. To specify the dominant source of a feeling of love (cortex or subcortex), we must ask ourselves what will happen if we no longer feel the pleasure of that love.

If our partner, spouse, parents, children, friend or an object becomes ill or requires repair, and the strength of our love for them is unchanged, we will come forward to help or try to fix it. This reaction shows that our love has a strong and dominant component originating from the cortex. If, instead, the intensity of our love decreases at such a time, perhaps to the point of indifference, then the dominant component of this love comes from the subcortex. An expression of love that is predominately from the cortex can be found in the following story.

After a lecture I gave on love, a man from the audience came up to me and said, "Dr Chalamish, I am 91 years old and it is important for me to tell you something. I married my wife when I was 28, and it wasn't for love, but that was how it often happened in those days. Today, my wife is 87 with advanced Alzheimer's disease, and she also suffers from a muscular condition that causes her to be bedridden." I told him that I was sorry to hear this. He continued, "Dr Chalamish, listen to me carefully. I love my wife today more than ever, and my love for her is only getting stronger and stronger."

Now that we have been exposed to the tip of the iceberg of the mental side of love, I would like to detail two forms of love with enormous potential for promoting physical and mental health: self-love and romantic love.

Self-love

Like the other forms of love, there are two potential paths that lead to self-love – one emanating from the subcortex and another from the cortex. However, unlike the other types of love, according to the brain code, self-love from the subcortex is innate. That is, we do not need to convince the subcortex to trigger the motivation for self-love to advance our survival. Self-love that originates from the subcortex will manifest as a strong drive for self-preservation, which is expressed as automatic behaviour to escape from danger and our ability to recognize opportunities to enhance pleasure. In situations of mental disruption, this self-love will wane and it follows that our natural reflexes to respond effectively to situations threatening our survival will also weaken. This can be a gradual process, starting from an early age. For example, adults who express suicidal thoughts will say that, from a young age, they were not interested in continuing to live.

Similarly, self-love from the subcortex might dissipate as a result of a mental illness, such as depression, a psychotic mental disorder or a processing system in the brain that fails to ensure someone adapts to a feeling of extreme suffering. (This last is a characteristic associated with those with a tendency to develop personality disorders.) In all these situations, being unable to apply self-love may lead to suicide.

The activity involved for the cortex to express self-love is more complex than the innate reflex on which this emotion is based when it originates from the subcortex. The cortex, as with the other forms of love, is not in a hurry to activate the neural network for self-love. The cortex requires concrete proof that the person deserves to feel intense self-love. This is an excellent reason to emphasize all the achievements in our day-to-day lives. Every time we work to nurture others, this will be expressed in the brain as contributing to building up our temple of self-love in the cortex.

The positive expression of self-love is an enhanced, healthy form of self-esteem originating from the cortex (as opposed to the narcissistic form of self-love, loving yourself and putting yourself before others). Studies note that this positive kind of self-love or self-esteem is one of the most important functions of the brain, which is also called the mental vaccination.[5] People with high levels of this self-love or self-esteem have improved immunity to mental health issues and deal effectively with situations of psychological stress and other medical problems.

It is important to distinguish between high levels of self-esteem, which appear as an activity of a neural network in the cortex for self-love, and increased self-confidence, which is merely expressed as a behavioural component related to self-esteem. These are two different brain functions that do not depend on each other. Therefore, there are people who demonstrate great self-confidence but suffer from low self-esteem. In such cases, the flamboyant behaviour (which might convince some that they are well-balanced individuals) has no solid foundation in true self-love. This way of being can be likened to a cheque that is made out for a huge sum of money but there are no funds in the account.

Such people can be dangerous because if it's only self-love originating from the subcortex that is driving their emotions and behaviour, then in life-threatening situations, for the sake of survival, they will ignore societal norms and the wellbeing of people surrounding them. We see such situations when people have narcissistic personality disorder and antisocial personalities.

Fortunately, at the opposite end of this scale, we find people who have well-founded high levels of self-esteem and self-confidence. These people experience enhanced mental wellbeing and are endowed with a wonderful capacity for empathy. Their self-love is projected out to their environment and instils a sense of security among the people around them. Such individuals will have good and creative means of dealing with threatening situations, and any potential harm to others will be considered before any response is made or action taken. To continue our analogy, these people can be likened to a cheque backed up by hard cash.

Romantic love

Romantic love can develop unilaterally or bilaterally and might be directed at someone of the opposite or same sex or someone who is lesbian, gay, bi, trans, queer, questioning or asexual (LGBTQ+). In general, romantic love is essentially no different from the love we feel for siblings, parents, children and friends. It, too, will be expressed in terms of feelings and urges for closeness with a loved one, and will be experienced in different intensities. When the intensity is too weak, there will be a tendency to feel indifference, and if the intensity is excessive, strong feelings will manifest in behaviour that is characteristic of an addiction or obsession. Romantic love, like the other forms of love, has two independent generators in the brain that drive it: one in the subcortex and the other in the cortex.

Now that we have noted the similarities between romantic and non-romantic love, we shall get to know the characteristics that are unique to the romantic kind. Let's begin with the romantic love that originates in the subcortex and then we will move "upstairs" to the cortex.

The role of the subcortex

The subcortex, which is our deep brain, is tuned to promoting our survival. Its role is to receive messages from the environment, process them and generate feelings and behavioural impulses that will benefit us right now or in the short term. The brain also has no problem assigning motivation to the subcortex if the outcome of the processing of messages promotes pleasure. If received messages are perceived as a threat to survival, the brain is quickly motivated to create an emotion of fear. This leads to automatic behaviour with electrical and chemical signals telling the body to get ready for a confrontation (see chapter 3 for more on this).

Pleasure sits at the opposite end of the spectrum for survival from a threat. It is a situation in which survival in the immediate future is certain. Consequently, there is a close connection in the subcortex between romantic love and the emotion of pleasure. According to the brain code, in common with other animals, we humans also deserve to have offspring. In fact, having them is an important part of our life's mission. To convince us that we should reproduce, the subcortex makes sex relevant to our survival by linking it to the arousal of emotions in which pleasure plays a dominant role.

Every person, from adolescence onward, has the potential for sexual desire. The manifestation of this is the subcortex arousing emotions toward a person who has passed the laboratory tests in our deep brain. Desire can develop simultaneously toward several people, and it is spread by the male and female sex hormones, testosterone and oestrogen. These hormones are only secreted from the age of sexual maturity, so prepubescent children don't feel sexual desire toward another child, though they can be excited by the curiosity of discovery.

Attraction is the first emotion generated by the subcortex that is an expression of romantic love that we would broadly define as sexuality. As previously mentioned, our ancient brain code attributes great importance to reproduction. Even when faced with food scarcity, fearsome beasts of prey, harsh nature and a precarious existence, the brain allocated resources to reproduction. The goal of survival here is clear: getting pregnant. The sophisticated laboratory in the subcortex receives massive amounts of data from the senses about potential mates. In many cases, it will give a "Go" signal to the further development of desire, but when it comes to falling in love, it's a completely different story.

Before giving the go-ahead to fall in love, the subcortex laboratory examines the potential partner in higher resolution, as it were. As far as we know, the subcortex does not run DNA tests but, thanks to the potential candidate's smell, touch, saliva, appearance and speech samples, the laboratory can create a profile of the candidate and determine whether they are a good match. Unlike other expressions of love, falling in love is a game on which the brain bets your entire stake. All in! Go or no go! Either you walk the flower-strewn path to your love story or you wander alone on a rocky trail. If the laboratory's diagnosis is positive, the outcome will be falling in love.

Two substances naturally secreted by the brain in response to romance are dopamine and serotonin. Dopamine plays many roles in the brain, but its function in relation to our topic here is the arousal of emotions to promote a feeling of pleasure. Dopamine causes the mind of a person in love to support behaviour that produces pleasure.

Serotonin, likewise, is a multifunctional brain chemical but, in relation to love, its role is in the higher-order functions of the cortex, such as memory, thinking and regulation of behaviour. When we fall in love, serotonin production and its effects are suppressed, so it follows that the system of rational alerts, which warn us about future scenarios that may damage our wellbeing, slows down. Therefore, without serotonin there is no responsible adult in the brain. It's party time!

The situation created by the suppression of the cortex and the sensuousness of the subcortex is that the brain of the person in love is like a runaway train, unable to consider what is happening around them. The feeling is euphoric. The intense emotional arousal fuelled by dopamine allows them to be active and vital in everything related to the implementation of closeness with the object of their desire. But sleep, food, friends and family become secondary during this period with the focus is on the object of their love. As mentioned, the cortex supports this situation and diverts from the conscious mind any unnecessary thoughts, such as, "I didn't sleep well last night", "I forgot my mother's birthday" or "My manager is out to get me." Furthermore, at this stage of falling in love, even if friends warn that the focus of this love has a dubious past or they have some other negative information about them, it's highly doubtful these messages will penetrate the sleepy cortex and make any difference.

This state of falling in love requires huge amounts of energy and for many of us it does not last long. Feelings of elation, vitality and pleasure are temporary. The brain allocates a period of up to several weeks for us to experience this wonderful feeling, but what happens when it wears off?

The subcortex will decide whether to continue the relationship, developing a more stable feeling of love toward our loved one, or revert to its original state. That is, shut down the focused emotional connection and move back to a state of general desire rather than this specific one. Studies have found a correlation between the kind of brain activity which occurs under the influence of addictive drugs and that observed when someone is in the state of falling in love. Similarly, breaking up with the object of their affection while the infatuation mechanism is still active is the same as detoxing from an addictive drug.[6] The mental pain is unbearable and, at the same time, there may also be physical symptoms. Over time and, thanks to the willpower we generally have to carry on, the relevant parts of the brain will undergo a physiological adaptation, leading to a significant relief in sensations and symptoms.

Should the relationship continue, attachment is the next, higher-level stage in the realization of love in the subcortex after attraction. Attachment feels quite different because emotional arousal is lowered from the high state of excitement of the initial attraction back down, closer to the normal base level, which gives stability to the feelings of romance. The warm feelings, the urge for intimacy and the desire for sexual contact are still present, but the flame has been turned down and held at a steady burn. The brain stores the feeling of falling in love in the memory, which enables intense, highly emotional romantic moments to be reconstructed so the couple can continue to enjoy them. In these moments (thanks to the stored memory and the motivation there is to restore it), the brain can bring back that special exciting feeling, even if it's only for a few hours.

The mental and behavioural components of attachment obviously have a physiological source, in the form of hormones. Their role is to activate the formation of unique neural networks in the subcortex that, in turn, allow the emotions, impulses and memories related to this stage of love in the subcortex to be activated.[7] Based on these findings, many neuroscientists believe that the brain code supports a monogamous lifestyle.[8]

The first of these hormones is oxytocin, which is secreted by the pituitary gland. This is located in the lower part of the brain, level with the top of the nose.[9] Oxytocin penetrates the small blood vessels, called capillaries, located close to the pituitary gland. From there, the oxytocin makes its way via the bloodstream throughout the body.

During the 20th century, scientific understanding of the functions of oxytocin were that it was to aid childbirth by stimulating contractions, control bleeding after birth, help with breastfeeding by aiding the let-down reflex, and foster closeness between a mother and her baby. It was only more recently, as part of the flexible brain revolution, that researchers came to understand that it also makes a tremendous contribution to emphasizing and shaping romantic love generated in the subcortex. Hence it is nicknamed the bonding hormone and the love hormone.[10] Physical closeness to a loved one will promote secretion of the hormone to strengthen the closeness.

The second hormone involved in the attachment phase is vasopressin. Like oxytocin, it is secreted by the pituitary gland into the bloodstream. Researchers formerly assumed that its function was limited to constricting blood vessels and preventing the excretion of water in the urine if the body was dehydrated, but neuroscientists have since recognized that it has an additional, completely different role.[11]

The brain wants to ensure that the person in love who progresses to the attachment stage will avoid feeling passionate toward other potential lovers, and this is where vasopressin steps in. It's nickname is the cheating prevention hormone. I'm sure many of you are now raising an eyebrow and wondering how there is a prevalence of cheating and extra-marital relationships. My response is that this comes from culture – the mind-shaping influence of culture. So in a cultural environment where cheating exists, this "new" knowledge may replace the ancient knowledge that exists in the form of the brain code.

The attachment phase may last a long time and manifest in the urge for intimacy, passion and the ability to recreate the delightful moments of falling in love from memories. Sounds good, doesn't it? But let's not forget that the source of this attachment is feelings generated by the subcortex, so it is all part of the brain's strategy to promote survival and not to improve our wellbeing.

Married partners who express their love mainly as attachment, without some measure of love being generated by the cortex, can expect to hit a few bumps. Life has a habit of throwing up problems, such as a depleted bank account, difficulties that come with having children and other disruptions to managing a household that may cause anxieties and cracks to appear in the relationship. This can be likened to a sunny day but then a cloud blocks the sun and a shadow forms. The negative effects of such situations will be few and far between if the love generated by the subcortex is balanced by love of a similar intensity that has originated from the cortex.

The role of the cortex

The cortex, as we learned in chapter 1, is an area of the brain that was added to the ancient subcortex in the process of our evolution. Fish and reptiles have no cortex and run their lives solely by means of their active subcortex, which navigates them to safe environments where they can find food, be protected from predators and reproduce. We humans are privileged. In addition to an active subcortex, we possess a developed cortex that enables us to be aware of ourselves and the physical and human environment that surrounds us. The cortex also allows us to process our thoughts and has endowed us with the gift of verbal communication. Thanks to this we can direct our lives to advance our mental wellbeing and not merely promote our immediate survival. The cortex has applied copy and paste actions to a wide range of the functions that happen in the subcortex, such as memory, movements, spatial orientation and vision. It has also given them a new purpose, which is to strive to achieve that elusive thing called happiness and promote our wellbeing.

The aim of romantic love in the subcortex is to ensure reproduction and the survival of our offspring. In the cortex, the goal of romantic love is the advancement of our wellbeing. And this extends beyond our own life, to that of our beloved. This advancement is mediated by emotional expressions and behaviour. So romantic love expressed by the cortex removes the ego from the relationship and introduces empathy. This is the ability to step into the shoes of the loved one, understand what is right for them at a particular moment and implement a behaviour that will support and empower them.

It is customary to describe the development of romantic love as first originating in the subcortex and only later also being present in the cortex. In many cases, it is indeed the process, but it is important to note that all variations are possible. Sometimes the flame of romantic love will be fuelled by the cortex instead, with the subcortex having the secondary role. Regarding the division of labour to create perfect romantic love, the intense, passionate moments will be the responsibility of the subcortex, while the times that call for an honest conversation, a feeling of closeness and intimacy, will be ruled by the cortex.

Remember my friend who was holding up the door frame at the beginning of this chapter? The romantic love he feels for his wife stems from both the subcortex *and* the cortex. For him, the love coming from these two parts of the brain enable him to express his love for her at all times, and they will overlap each other when needed. When the clouds hide the sun and life events overshadow the desire that comes from his subcortex, my friend is able to concentrate on and emphasize the intimate friendship that he also loves and so keep his love for her alive. Sharing his wife's difficulties and the empathy he will show during their soul talks will strengthen their relationship and let them cooperate to find solutions.

Some tips for improving your self-love

As we have seen, self-esteem is the positive expression of self-love. Studies reveal that a person endowed with high self-esteem will have increased immunity to mental and physical problems such as anxiety and inflammation. So how can your self-love and, therefore, self-esteem be improved?

1 Think of situations in which you have felt self-esteem. This might be winning a competition, scoring highly in a test, success at work or an achievement in the family.
2 To ensure that this feeling of self-esteem is encoded in the brain, write down the details of the event and record yourself reading what you have written.
3 Listen to the recording each day, concentrating on the words, and allow yourself to absorb that feeling of self-esteem, just as you enjoyed feeling it in the moment when it happened.
4 As you go about your day, emphasize to yourself each small achievement as it occurs. Cultivate a happy and optimistic expectation that there will be other achievements yet to come, and look forward to them.

***

Returning to the analogy of the sun and the clouds of my friend's relationship with his wife, the most beautiful sunrises and sunsets occur when clouds hide the sun because then the sky glows a wonderful crimson.

EPILOGUE

We have now concluded our journey around the human brain. I hope you have enjoyed it and come away with a feeling of admiration for this wonderful part of ourselves that is at the core of our existence. I also hope I have been able to dispel the concept that the functions of the brain and behaviour are beyond your control, and that it is now clear this is not true. With the help of awareness, motivation and proper practices and exercises you have now learned about, any problem areas can be tackled and even eliminated, including those which cause suffering and harm in our day-to-day lives.

For the past ten years, in collaboration with Ramon Velleman, I have been promoting an inclusive conversation – for everyone, not only professionals – about the extraordinary capacity of the brain to improve and even correct how it performs. We have been spreading the word, through lectures and in digital training programmes, about how to improve your memory and the other incredible abilities your brain has and can develop. You can find these in my Do4Brain app, which allows you to exercise your brain, whether for pleasure or out of curiosity, on a daily basis. I invite you to continue exploring and learning about the brain with me.

A WORD OF THANKS

I would like to thank the following people who contributed significantly to the writing and publishing of this book.

- Ramon Velleman for his support, friendship and wonderful, life-shaping advice
- Yosef Shavit, who managed to transcribe all our fascinating conversations
- Amit Hadad and Topaz Lushi Kay, who with rare skill managed to track down and identify the scientific sources for all the research mentioned in the book
- Yuval Elazari, for his warm and intelligent attitude while editing
- the staff at Kinneret Zamora Publishing for their personal and professional attitude while publishing the original Hebrew version of this book

Some teachers have been significant in my life as I have also learned from them that simplicity, modesty and love for people are more important than simply accumulating knowledge. I am grateful to:

- the late Ofra Avidor
- the late Yitzhak Noy
- the late Avihu Yeshurun
- the late Arye Nir
- Amos Etzioni
- Michael Kaufman
- Rodika Goychman
- Ilana Kremer
- Lior Fish
- Yadin Dodai
- Alexander Solomonovich
- Dani Kerman

I also want to express my gratitude to Tsiyona Peled, my friend and tutor, for the wise advice that helped me to improve the accuracy of my writing.

Finally, thanks to my family for their encouragement and faith in me, to my brother, Zvi Chalamish, and the Foundation Panama-Federico for their generous help, and to all my friends, teachers, students and patients, from whom I have learned so much.

NOTES

1 A Brief History of the Brain

1   Doidge, N, *The Brain that Changes Itself: Stories of personal triumph from the frontiers of brain science*, Penguin, London, 2008

2 Understanding Memory

1   Roediger III, H L, Dudai, Y and Fitzpatrick, S M (eds), *Science of Memory: Concepts*, Oxford University Press, Oxford, 2007

2   De la Fuente, I M, Bringas, C, Malaina, I, et al., "Evidence of conditioned behavior in amoebae", *Nature Communications*, 2019, 10(1): 3690

3   Ueda, T, Matsumoto, K and Kobatake, Y, "Perception in an amoeboid cell", in Mishra R K (ed.), *Molecular and Biological Physics of Living Systems*, Springer, Dordrecht, 1990, 133–145

4   Eagleman, D, *Incognito (Enhanced Edition): The secret lives of the brain*, Knopf Doubleday, New York, 2011

5   Monteiro, C A, Moubarac, J-C, Cannon, G, Ng, S W and Popkin, B, "Ultra-processed products are becoming dominant in the global food system", *Obesity Reviews*, 2013, 14(2): 21–28

6   Kokubo, Y, Higashiyama, A, Watanabe, M and Miyamoto, Y, "A comprehensive policy for reducing sugar beverages for healthy life extension", *Environmental Health and Preventive Medicine*, 2019, 24(1): 1–4

7   Elvsåshagen, T, Norbom, L B, Pedersen, P Ø, et al., "Widespread changes in white matter microstructure after a day of waking and sleep deprivation", *PLOS ONE*, 2015, 10 (5): e0127351; Liu, C, Kong, X-Z, Liu, X, et al., "Long-term total sleep deprivation reduces thalamic gray matter volume in healthy men", *Neuroreport*, 2014, 25(5): 320–323

8   Oei, N Y L, Everaerd, W T A M, Elzinga, B M, Well, S van and Bermond, B, "Psychosocial stress impairs working memory at high loads: An association with cortisol levels and memory retrieval", *Stress*, 2006, 9(3): 133–141;

Wolf, O T, "Stress and memory in humans: Twelve years of progress?", *Brain Research*, 2009, 1293:142–154

9   Roig, M, Nordbrandt, S, Geertsen, S S and Nielsen, J B, "The effects of cardiovascular exercise on human memory: A review with meta-analysis", *Neuroscience and Biobehavioral Reviews*, 2013, 37(8): 1645–1666

10 Lieberman, D E, "Is exercise really medicine?: An evolutionary perspective", *Current Sports Medicine Reports*, 2015, 14(4): 313–319

11 Pontzer, H, Wood, B M and Raichlen, D A, "Hunter-gatherers as models in public health", *Obesity Reviews*, 2018, 19(1): 24–35

12 Frith, C, *Making up the Mind: How the brain creates our mental world*, Wiley-Blackwell, Hoboken, NJ, 2007

13 Loftus, E F and Pickrell, J E, "The formation of false memories", *Psychiatric Annals*, 1995, 25(12): 720–725

14 The research was undertaken in collaboration with Rachel Ludmer and Yadin Dudai from the Weizmann Institute, Israel, as part of Ludmer's doctoral thesis on brain research. See Ludmer, R, "Brain correlates of encoding, modification and recollection of human one shot learning", PhD thesis, Weizmann Institute of Science, Israel, 2012

15 Mendelsohn, A, Chalamish, Y, Solomonovich, A and Dudai, Y, "Mesmerizing memories: Brain substrates of episodic memory suppression in posthypnotic amnesia", *Neuron*, 2008, 57(1): 159–170

3 Emotion Reigns Supreme

1   Luo, J and Yu, R "Follow the heart or the head?: The interactive influence model of emotion and cognition", *Frontiers in Psychology*, 2015, 6: 573

2   Pileggi, T, "Banker who embezzled NIS 250 million wins early release", *Times of Israel*, 2023, 4 December. Available at: www.timesofisrael.com/banker-who-embezzled-nis-250-million-wins-early-release (accessed November 2023)

3   Frith, C, *Making up the Mind: How the brain creates our mental world*, Wiley-Blackwell, Hoboken, NJ, 2007

4   Hooker, S A, Masters, K S and Park, C L, "A meaningful life is a healthy life: A conceptual model linking meaning and meaning salience to health", *Review of General Psychology*, 22(1): 11–24

5   Based on ideas in D Pink's book *Drive: The surprising truth about what motivates us* (Canongate Books, 2018) and a TED talk he gave, "The puzzle of motivation", July 2009. Available at: www.ted.com/talks/dan_pink_the_puzzle_of_motivation (accessed November 2023)

6   Frankl, V, *Man's Search for Meaning*, Rider, 2004

4 The Colours of Emotion

1   Ford, B Q, "Anger gives you a creative boost", *Scientific American*, 2011, 23 August; Ratson, M, "The value of anger: 16 reasons it's good to get angry", *GoodTherapy* blog, 2017, 13 March. Available at: www.goodtherapy.org/blog/value-of-anger-16-reasons-its-good-to-get-angry-0313175 (accessed November 2023)

5 Struggles of the Mind

1   Liu, Q, He, H, Yang, J, et al., "Changes in the global burden of depression from 1990 to 2017: Findings from the global burden of disease study", *Journal of Psychiatric Research*, 2020, 126: 134–140
2   Hagen, E H, "The functions of postpartum depression", *Evolution and Human Behavior*, 1999, 20(5): 325–359
3   Pugh, J and Ohler, N, "Blitzed: Drugs in Nazi Germany", *British Journal for Military History*, 2017, 3: 160–162
4   Pugh and Ohler, "Blitzed: Drugs in Nazi Germany"

6 Our Wonderful Senses

1   Suzuki, M, Pennartz, C M and Aru, J, "How deep is the brain?: The shallow brain hypothesis", *Nature Reviews Neuroscience*, 2023, 24(12): 1–14
2   Herbet, G and Duffau, H, "Revisiting the functional anatomy of the human brain: Toward a meta-networking theory of cerebral functions", *Physiological Reviews*, 2020, 100(3): 1181–1228
3   Celeghin, A, Bagnis, A, Diano, M, et al., "Functional neuroanatomy of blindsight revealed by activation likelihood estimation: Meta-analysis", *Neuropsychologia*, 2019, 128:109–118
4   Laan, L N van der, Ridder, D T de, Viergever, M A and Smeets, P A M, "The first taste is always with the eyes: A meta-analysis on the neural correlates of processing visual food cues", *Neuroimage*, 2011, 55(1): 296–303
5   PresidentialConf, "It's all in your head: How do we really form opinions? – Professor Noam Sobel", YouTube, 2013, 20 June. Available at: www.youtube.com/watch?v=THVEVv9LvYU (accessed November 2023)
6   McClintock, M K, "Menstrual synchrony and suppression", *Nature*, 1971, 229(5282): 244–245
7   Gosline, A, "Do women who live together menstruate together?", *Scientific American*, 2007. Available at: www.scientificamerican.com/article/do-women-who-live-together-menstruate-together (accessed November 2023)

8   Gelstein, S, Yeshurun, Y, Rozenkrantz, L, et al., "Human tears contain a chemo-signal", *Science*, 2011, 331(6014): 226–230

9   Frumin, I, Perl, O, Endevelt-Shapira, Y, et al., "A social chemosignaling function for human handshaking", *Elife*, 2015, 4: e05154

10 Zaraska, M, "The sense of smell in humans is more powerful than we think", *Discover*, 2017, 17 April 2020. Available at: www.discovermagazine.com/mind/the-sense-of-smell-in-humans-is-more-powerful-than-we-think (accessed November 2023)

11 Williams, L E and Bargh, J A, "Experiencing physical warmth promotes interpersonal warmth", *Science*, 2008, 322(5901): 606–607

12 Ackerman, J M, Nocera, C C and Bargh, J A, "Incidental haptic sensations influence social judgments and decisions", *Science*, 2010, 328(5986): 1712–1715

13 Hertenstein, M J, Keltner, D, App, B, Bulleit, B A and Jaskolka, A R, "Touch communicates distinct emotions", *Emotion*, 2006, 6(3): 528–533

14 Field, T M, "Interventions for premature infants", *Journal of Pediatrics*, 1986, 109(1): 183–191

15 Ang, J Y, Lua, J L, Mathur, A, et al., "A randomized placebo-controlled trial of massage therapy on the immune system of preterm infants", *Pediatrics*, 2012, 130(6): e1549–e1558

16 Crusco, A H and Wetzel, C G, "The Midas touch: The effects of interpersonal touch on restaurant tipping", *Personality and Social Psychology Bulletin*, 1984, 10(4): 512–517

17 Nagasako, E M, Oaklander, A L and Dworkin, R H, "Congenital insensitivity to pain: An update", *Pain*, 2003, 101(3): 213–219

18 Higgins, D M, Martin, A M, Baker, D G, et al., "The relationship between chronic pain and neurocognitive function: A systematic review", *Clinical Journal of Pain*, 2018, 34(3): 262–275

19 Henry, D E, Chiodo, A E and Yang, W, "Central nervous system reorganization in a variety of chronic pain states: A review", *Physical Medicine and Rehabilitation and Acute Inpatient Rehabilitation*, 2011, 3(12): 1116–1125

20 Hillier, S and Worley, A, "The effectiveness of the Feldenkrais method: A systematic review of the evidence", *Evidence-Based Complementary and Alternative Medicine*, 2015, Article ID: 752160

21 Clearfield, M W, "Learning to walk changes infants' social interactions", *Infant Behavior and Development*, 2011, 34(1): 15–25

22 Ramachandran, V S and Hubbard, E M, "Synaesthesia: A window into perception, thought and language", *Journal of Consciousness Studies*, 2001, 8(12): 3–34. Available at: www.sfu.ca/~kathleea/colour/docs/Ram%26Hub_2001.pdf (accessed November 2023)

23 Ramachandran, V S and Hubbard, E M, "Synaesthesia: A window into perception, thought and language", *Journal of Consciousness Studies*, 2001, 8(12): 3–34

24 E Gerti, "Synaesthesia: The ability to smell colours", Weizmann Institute, 2010, 28 June. Available at: www.ncbi.nlm.nih.gov/pmc/articles/PMC3222625/ (accessed November 2023)

7 How the Brain Creates Reality

1 Purepedantry, "Intact visual navigation in a patient with blindsight", YouTube, 2009. Available at: www.youtube.com/watch?v=nFJvXNGJsws (accessed November 2023)

2 Eagleman, D, *Incognito (Enhanced Edition): The secret lives of the brain*, Knopf Doubleday, New York, 2011

3 Simons, D, "Selective attention test: From Simons and Chabris (1999)", YouTube, 2010. Available at: www.youtube.com/watch?v=vJG698U2Mvo (accessed November 2023)

4 Simons, D, "The monkey business illusion", YouTube, 2020. Available at: www.youtube.com/watch?v=IGQmdoK_ZfY (accessed November 2023)

5 Frith, C, *Making up the Mind: How the brain creates our mental world*, Wiley-Blackwell, Hoboken, NJ, 2007

6 RayOman, "Charlie Chaplin optic [sic] illusion", YouTube, 2006. Available at: www.youtube.com/watch?v=QbKw0_v2clo 106 (accessed November 2023)

7 Goldberg, I I, Harel, M and Malach, R, "When the brain loses its self: Prefrontal inactivation during sensorimotor processing", *Neuron*, 2006, 50(2): 329–339

8 Dengler, R, "What causes hallucinations?: The brain may be overinterpreting a lack of info", *Discover*, 2019, 27 March. Available at: www.discovermagazine.com/mind/what-causes-hallucinations-the-brain-may-be-overinterpreting-a-lack-of-info (accessed November 2023)

9 Mendelsohn, A, Chalamish, Y, Solomonovich, A and Dudai, Y, "Mesmerizing memories: Brain substrates of episodic memory suppression in posthypnotic amnesia", *Neuron*, 2008, 57(1): 159–170

10 Mischel, W, Shoda, Y and Ayduk, O, *Introduction to Personality: Toward an integrative science of the person*, John Wiley & Sons, Hoboken, NJ, 2007

11 Kahneman, D, *Thinking, Fast and Slow*, Farrar, Straus & Giroux, New York, 2011

12 Shleifer, A, "Psychologists at the gate: A review of Daniel Kahneman's *Thinking, Fast and Slow*", *Journal of Economic Literature*, 2012, 50(4): 1080–1091

13 Kraft, U, "Unleashing creativity", *Scientific American Mind*, 2005, 16(1): 16–23

14 Seelig, T, *Insight Out: Get ideas out of your head and into the world*, Harper-One, San Francisco, CA, 2015

15 Kerman told me this story during a break at one of these joint workshops

8 The Learning Brain

1 Kahneman, D, *Thinking, Fast and Slow*, Farrar, Straus & Giroux, New York, 2011

2 Kahneman, *Thinking, Fast and Slow*

3 Mischel, W, *The Marshmallow Test: Why self-control is the engine of success*, Little, Brown Spark, New York, 2014

4 Mischel, *The Marshmallow Test*

5 Bloom, B S, Engelhart, M, Furst, E J, Hill, W and Krathwohl, D R, *Taxonomy of Educational Objectives, Handbook I: Cognitive domain*, Longman, New York, 1956

9 Our Relationship with the World Shapes Our Personality

1 Whalen, P J, Rauch, S L, Etcoff, N L, et al., "Masked presentations of emotional facial expressions modulate amygdala activity without explicit knowledge", *Journal of Neuroscience*, 1998, 18(1): 411–418

2 Hammond, D C, "Hypnosis as sole anaesthesia for major surgeries: Historical and contemporary perspectives", *American Journal of Clinical Hypnosis*, 2013, 51(2): 101–121

3 Jansen, A S, Nguyen, X V, Karpitskiy, V, Mettenleiter, T C and Loewy, A D, "Central command neurons of the sympathetic nervous system: Basis of the fight-or-flight response", *Science*, 1995, 270(5236): 644–646

10 Taking Care of Our Grey Matter

1 Feltz, D L, "Self-confidence and sports performance", in D Smith and M Bar-Eli (eds), *Essential Readings in Sport and Exercise Psychology*, Human Kinetics, Champaign, IL, 2007, pp. 278–294

2 Daly, S, Thorpe, M, Rockswold, S., et al., "Hyperbaric oxygen therapy in the treatment of acute severe traumatic brain injury: A systematic review", *Journal of Neurotrauma*, 2018, 35(4): 623–629

3 Vadas, D, Kalichman, L, Hadanny, A and Efrati, S, "Hyperbaric oxygen environment can enhance brain activity and multitasking performance", *Frontiers in Integrative Neuroscience*, 2017, 11: 25

4   The Nobel Prize in Physiology or Medicine 1986. Available at: www.nobel-prize.org/prizes/medicine/1986/summary (accessed November 2023)

5   Mintzer, J, Donovan, K A, Kindy, A Z, et al., "Lifestyle choices and brain health", *Frontiers in Medicine*, 2019, 6: 204

6   Robbins, R, Grandner, M A, Buxton, O M, et al., "Sleep myths: An expert-led study to identify false beliefs about sleep that impinge upon population sleep health practices", *Sleep Health*, 2019, 5(4): 409–417

7   Everson, C A, Bergmann, B M and Rechtschaffen, A, "Sleep deprivation in the rat: III: Total sleep deprivation", *Sleep*, 1989, 12(1): 13–21

8   Irwin, M R, Olmstead, R and Carroll, J E, "Sleep disturbance, sleep duration, and inflammation: A systematic review and meta-analysis of cohort studies and experimental sleep deprivation", *Biological Psychiatry*, 2016, 80(1): 40–52; Pires, G N, Bezerra, A G, Tufik, S and Andersen, M L, "Effects of acute sleep deprivation on state anxiety levels: A systematic review and meta-analysis", *Sleep Medicine*, 2016, 24: 109–118; Zhai, L, Zhang, H and Zhang, D, "Sleep duration and depression among adults: A meta-analysis of prospective studies", *Depression and Anxiety*, 2015, 32(9): 664–670

9   Lin, X, Chen, W, Wei, F, et al., "Night-shift work increases morbidity of breast cancer and all-cause mortality: A meta-analysis of 16 prospective cohort studies", *Sleep Medicine*, 2015, 16(11): 1381–1387

10 Chaput, J P, Dutil, C, Featherstone, R, et al., "Sleep duration and health in adults: An overview of systematic reviews", *Applied Physiology, Nutrition, and Metabolism*, 2020, 45(10): S218–S231

11 Ma, Y, Liang, L, Zheng, F, et al., "Association between sleep duration and cognitive decline", *JAMA Network Open*, 2020, 3(9): e2013573

12 White, A J, Weinberg, C R, Park, Y M, et al., "Sleep characteristics, light at night and breast cancer risk in a prospective cohort", *International Journal of Cancer*, 2017, 141(11): 2204–2214

13 Olaithe, M, Bucks, R S, Hillman, D R and Eastwood, P R, "Cognitive deficits in obstructive sleep apnea: Insights from a meta-review and comparison with deficits observed in COPD, insomnia, and sleep deprivation", *Sleep Medicine Reviews*, 2018, 38: 39–49

11 Caution: Dangerous Food

1   Jabr, F, "How sugar and fat trick the brain into wanting more food", *Scientific American*, 2016, 1 January

2   Menotti, A and Puddu, P E, "How the 'Seven countries study' contributed to the definition and development of the Mediterranean diet concept: A 50-year

journey", *Nutrition, Metabolism and Cardiovascular Diseases*, 2015, 25(3): 245–252

3   Imatome-Yun, N, "Bad science or bad journalism?: Top experts come together to address nutrition myths", Blue Zones. Available at: www.bluezones.com/2017/08/top-experts-come-together-to-address-nutrition-myths (accessed November 2023). See also Pett, K D, Kahn, J, Willett, W C and Katz, D L, "Ancel Keys and the Seven countries study: An evidence-based response to revisionist histories", White Paper, True Health Initiative, 2017, 1 August. Available at: www.truehealthinitiative.org/wp-content/uploads/2017/07/SCS-White-Paper.THI_.8-1-17.pdf (accessed November 2023)

4   Park, S, Ahn, J and Lee, B K, "Very-low-fat diets may be associated with increased risk of metabolic syndrome in the adult population", *Clinical Nutrition*, 2016, 35(5): 1159–1167

5   Teicholz, N, "A short history of saturated fat: The making and unmaking of a scientific consensus", *Current Opinion in Endocrinology, Diabetes and Obesity*, 2023, 30(1): 65–71

6   Yudkin, J, *Pure, White and Deadly: How sugar is killing us and what we can do to stop it*, Penguin Life, London, 2016

7   Ahmed, S H, Avena, N M, Berridge, K C, Gearhardt, A N, and Guillem, K, "Food addiction", in D W Pfaff, N D Volkow and J L Rubinstein (eds), *Neuroscience in the 21st Century: From basic to clinical*, Springer, Cham, pp. 4193–4218

8   Ahmed et al., "Food addiction"

9   Nakamura, E, "One hundred years since the discovery of the 'umami' taste from seaweed broth by Kikunae Ikeda, who transcended his time", *Chemistry: An Asian Journal*, 2011, 6(7): 1659–1663

10 Keast, R S and Costanzo, A, "Is fat the sixth taste primary?: Evidence and implications", *Flavour*, 2015, 4(5): 1–7

11 Myers, M G, Cowley, M A and Münzberg, H, "Mechanisms of leptin action and leptin resistance", *Annual Review of Physiology*, 2008, 70(1): 537–556

12 Harington, K, Smeele, R, Van Loon, F, et al., "Desire for sweet taste unchanged after eating: Evidence of a dessert mentality?", *Journal of the American College of Nutrition*, 2016, 35(6): 581–586

13 Keski-Rahkonen, A, "Epidemiology of binge eating disorder: Prevalence, course, comorbidity, and risk factors", *Current Opinion in Psychiatry*, 2021, 34(6): 525–531

14 Bernstein, B E, "Anorexia nervosa", Medscape, 2023, 22 June. Available at: emedicine.medscape.com/article/912187-overview#a6?form=fpf (accessed November 2023)

12 Fears and Anxieties

1   Pegrum, J and Pearce, O, "A stressful job: Are surgeons psychopaths?", *Bulletin of the Royal College of Surgeons of England*, 2015, 97(8): 331–334

2   Weinberger, D R, Elvevåg, B and Giedd, J N, "The adolescent brain: A work in progress", National Campaign to Prevent Teen Pregnancy, Washington, DC, June 2005

3   Fredrikson, M, Annas, P, Fischer, H and Wik, G, "Gender and age differences in the prevalence of specific fears and phobias", *Behaviour Research and Therapy*, 1996, 34(1): 33–39

4   Coelho, C M and Purkis, H, "The origins of specific phobias: Influential theories and current perspectives", *Review of General Psychology*, 2009, 13(4): 335–348

13 The Highest Level of the Brain's Functioning

1   Hunt, A W, Turner, G R, Polatajko, H, Bottari, C and Dawson, D R, "Executive function, self-regulation and attribution in acquired brain injury: A scoping review", *Neuropsychological Rehabilitation*, 2013, 23(6): 914–932

2   Bell, T E, "Robots in the home: Promises, promises: While great expectations are held for certain robot types, the robots for fun and educational purposes are limited in their adaptability to useful tasks", *IEEE Spectrum*, 1985, 22(5): 51–55; Stanger, C A, Anglin, C, Harwin, W S and Romilly, D P, "Devices for assisting manipulation: A summary of user task priorities", *IEEE Transactions on Rehabilitation Engineering*, 1994, 2(4): 256–265

3   Martinez-Conde, S and Macknik, S, *Champions of Illusion: The science behind mind-boggling images and mystifying brain puzzles*, Scientific American/Farrar, Straus & Giroux, New York, 2017

4   Perner, J, Frith, U, Leslie, A M and Leekam, S R, "Exploration of the autistic child's theory of mind: Knowledge, belief, and communication", *Child Development*, 1989, 60(3): 689–700

5   Visser, S N, Bitsko, R H, Danielson, M L, Perou, R and Blumberg, S J, "Increasing prevalence of parent-reported attention-deficit/hyperactivity disorder among children: United States, 2003 and 2007", *Morbidity and Mortality Weekly Report*, 2010, 59(44): 1439–1443

6   Nimmo-Smith, V, Merwood, A, Hank, D, et al., "Non-pharmacological interventions for adult ADHD: A systematic review", *Psychological Medicine*, 2020, 50(4): 529–541

7 Shelley-Tremblay, J F and Rosén, L A, "Attention deficit hyperactivity disorder: An evolutionary perspective", *Journal of Genetic Psychology*, 1996,
   157(4): 443–453

14 The Healing Brain

1 Dienes, Z, "Is hypnotic responding the strategic relinquishment of metacognition?", in M J Beran, J L Brandl, J Perner and J Proust (eds), *Foundations of
   Metacognition*, Oxford University Press, Oxford, 2012, pp. 267–278
2 Segnan, N, Minozzi, S, Armaroli, P, et al., "Epidemiologic evidence of slow
   growing, nonprogressive or regressive breast cancer: A systematic review",
   International Journal of Cancer, 2016, 139(3): 554–573. Available at: onlineli-
   brary.wiley.com/doi/pdf/10.1002/ijc.30105 (accessed November 2023)
3 Gilsinan, K, "The Buddhist and the neuroscientist: What compassion does to
   the brain", The Atlantic, 2015, 4 July. Available at: www.theatlantic.com/health/
   archive/2015/07/dalai-lama-neuroscience-compassion/397706/?utm_
   source=SFFB (accessed November 2023)
4 McGinn, L K and Sanderson, W C, "What allows cognitive behavioral therapy to be brief: Overview, efficacy, and crucial factors facilitating brief treatment", *Clinical Psychology: Science and Practice*, 2001, 8(1): 23–37
5 Definition of the word "hypnosis", the Free Dictionary by Farlex. Available at:
   www.thefreedictionary.com/hypnosis (accessed November 2023)
6 Reich, A, "Why is Israel's hypnosis law so strict?", *Jerusalem Post*, 2021, 27
   August. Available at: www.jpost.com/israel-news/why-is-israels-hypnosis-
   law-so-strict-677833 (accessed November 2023)
7 Levine, J and Salganik, I, "Conversation 14: Understanding hypnosis in the
   context of the internalized figures", Professor Joseph Levine's blog, 2022, 28
   October. Available at: joseph-levine.co.il/2022/10/understanding-hypnosis
   (accessed November 2023)
8 Kaptchuk, T J, Friedlander, E, Kelley, J M, et al., "Placebos without decep-
   tion: A randomized controlled trial in irritable bowel syndrome", *PLOS ONE*,
   2010,5(12): e15591
9 Willis, M T, "Knee surgery no better than placebo", ABC News, 2006, 10
   July. Available at: abcnews.go.com/%20Health/story?id=116879&page=1
   (accessed November 2023)
10 Amanzio, M, Benedetti, F, Porro, C A, Palermo, S and Cauda, F, "Activation
   likelihood estimation meta-analysis of brain correlates of placebo analgesia in
   human experimental pain", *Human Brain Mapping*, 2013, 34(3): 738–752

15 With Love in Mind

1   Berscheid, E, Dion, K, Walster, E and Walster, G W, "Physical attractiveness and dating choice: A test of the matching hypothesis", *Journal of Experimental Social Psychology*, 1971, 7(2): 173–189; Sharot, T, *The Influential Mind: What the brain reveals about our power to change others*, Henry Holt and Company, New York, 2017

2   Prinz, J J, *Beyond Human Nature: How culture and experience shape the human mind*, W W Norton & Company, New York, 2014

3   Lewandowsky, S, "The 'post-truth' world, misinformation, and information literacy: A perspective from cognitive science", in Goldstein, S (ed.), *Informed Societies: Why information literacy matters for citizenship, participation and democracy*, Facet, London, 2020, pp. 69–88; Oxford Languages, "Word of the Year 2016", Oxford University Press, 2016. Available at: languages.oup.com/word-of-the-year/2016 (accessed November 2023)

4   Peters, M A, "Education in a post-truth world", *Educational Philosophy and Theory*, 2017, 49(6): 563–566. Available at: www.tandfonline.com/doi/full/10.1080/00131857.2016.1264114 (accessed November 2023)

5   Campbell, W K, Rudich, E A and Sedikides, C, "Narcissism, self-esteem, and the positivity of self-views: Two portraits of self-love", *Personality and Social Psychology Bulletin*, 2002, 28(3): 358–368; Collins, A F, Turner, G and Condor, S, "A history of self-esteem: From a just honoring to a social vaccine", in D McCallum (ed.), *The Palgrave Handbook of the History of Human Sciences*, Palgrave Macmillan, Singapore, 2002, pp. 1117–1143

6   Cheng, Y, Chen, C, Lin, C P, Chou, K H and Decety, J, "Love hurts: An fMRI study", *Neuroimage*, 2010, 51(2): 923–929

7   Carter, C S, "The role of oxytocin and vasopressin in attachment", *Psycho dynamic Psychiatry*, 2017, 45(4): 499–517

8   Lu, Q, Lai, J, Du, Y, et al., "Sexual dimorphism of oxytocin and vasopressin in social cognition and behavior", *Psychology Research and Behavior Management*, 2019, 12: 337–349

9   Carson, D S, Guastella, A J, Taylor, E R and McGregor, I S, "A brief history of oxytocin and its role in modulating psychostimulant effects", *Journal of Psycho pharmacology*, 2013, 27(3): 231–247

10 Algoe, S B, Kurtz, L E and Grewen, K, "Oxytocin and social bonds: The role of oxytocin in perceptions of romantic partners' bonding behavior", *Psychological Science*, 2017, 28(12): 1763–1772; Colaianni, G, Sun, L, Zaidi, M and Zallone, A, "The 'love hormone' oxytocin regulates the loss and gain of the fat–bone relationship", *Frontiers in Endocrinology*, 2015, 6: 79

11 Rotondo, F, Butz, H, Syro, L V, et al., "Arginine vasopressin (AVP): A review of its historical perspectives, current research and multifunctional role in the hypothalamo-hypophysial system", *Pituitary*, 2016, 19(4): 345–355